WEYERHAEUSER ENVIRONMENTAL BOOKS

Paul S. Sutter, Editor

WEYERHAEUSER ENVIRONMENTAL BOOKS explore human relationships with natural environments in all their variety and complexity. They seek to cast new light on the ways that natural systems affect human communities, the ways that people affect the environments of which they are a part, and the ways that different cultural conceptions of nature profoundly shape our sense of the world around us. A complete list of the books in the series appears at the end of this book.

WETLANDS IN A DRY LAND

*More-Than-Human Histories of Australia's
Murray-Darling Basin*

EMILY O'GORMAN

UNIVERSITY OF WASHINGTON PRESS
Seattle

Wetlands in a Dry Land is published with the assistance of a grant from the Weyerhaeuser Environmental Books Endowment, established by the Weyerhaeuser Company Foundation, members of the Weyerhaeuser family, and Janet and Jack Creighton.

Copyright © 2021 by the University of Washington Press

Composed in Minion Pro, typeface designed by Robert Slimbach

25 24 23 22 21   5 4 3 2 1

Printed and bound in the United States of America

All rights reserved. No part of this publication may be reproduced or transmitted in any form or by any means, electronic or mechanical, including photocopy, recording, or any information storage or retrieval system, without permission in writing from the publisher.

UNIVERSITY OF WASHINGTON PRESS
uwapress.uw.edu

LIBRARY OF CONGRESS CATALOGING-IN-PUBLICATION DATA
Names: O'Gorman, Emily, author.
Title: Wetlands in a dry land : more-than-human histories of Australia's Murray-Darling Basin / Emily O'Gorman.
Description: Seattle : University of Washington Press, 2021. | Series: Weyerhaeuser environmental books | Includes bibliographical references and index.
Identifiers: LCCN 2020055835 (print) | LCCN 2020055836 (ebook) | ISBN 9780295749037 (hardcover) | ISBN 9780295749150 (paperback) | ISBN 9780295749044 (ebook)
Subjects: LCSH: Human ecology—Australia—Murray River Watershed (N.S.W.-S.A.) | Human ecology—Australia—Darling River Watershed (Qld. and N.S.W.) | Nature—Effect of human beings on—Australia—Murray River Watershed (N.S.W.-S.A.) | Nature—Effect of human beings on—Australia—Darling River Watershed (Qld. and N.S.W.) | Wetland management—Australia—Murray River Watershed (N.S.W.-S.A.) | Wetland management—Australia—Darling River Watershed (Qld. and N.S.W.) | Murray River Watershed (N.S.W.-S.A.)—Environmental conditions. | Darling River Watershed (Qld. and N.S.W.)—Environmental conditions.
Classification: LCC GF802.M87 O46 2021 (print) | LCC GF802.M87 (ebook) | DDC 304.2/509944—dc23
LC record available at https://lccn.loc.gov/2020055835
LC ebook record available at https://lccn.loc.gov/2020055836

The paper used in this publication is acid free and meets the minimum requirements of American National Standard for Information Sciences—Permanence of Paper for Printed Library Materials, ANSI Z39.48–1984.∞

This book was researched on and specifically discusses Wailwan, Wiradjuri, Ngarrindjeri, and Giabal Country. I am indebted to the generosity of Elders and custodians who shared knowledge of their Country and experiences with me. I also researched and wrote parts of this book on Gomeroi/Kamilaroi, Dharug, Gundungurra, Ngunnawal, Jagera, Boon Wurrung, and Woiwurrung Country. I further acknowledge Aboriginal custodians throughout the Murray-Darling Basin and Australia. I pay my respects to Elders, past, present, and emerging.

CONTENTS

Foreword: Entangled Agencies, by Paul S. Sutter   ix
Acknowledgments   xv
List of Abbreviations   xvii

INTRODUCTION   3

ONE.   Weaving: Postcolonial and Multispecies Politics of Plants   23

TWO.   Leaking: Containment and Recalcitrance of Swamps   46

THREE.   Infecting: Irrigation, Mosquitoes, and Malaria in Wartime   73

FOUR.   Crossing: Wildlife in Agriculture   97

FIVE.   Enclosing: Pelicans, Protected Areas, and Private Property   120

SIX.   Migrating: Wetlands, Transcontinental Bird Movements, and Global Environmental Crisis   141

SEVEN.   Rippling: Capitalism, Seals, and Baselines   168

AFTERWORD   194

Notes   199
Selected Bibliography   243
Index   251

FOREWORD

Entangled Agencies

PAUL S. SUTTER

In central Kansas, not far from the great bend in the Arkansas River, there sits one of the most prolific wetlands in North America. Cheyenne Bottoms Wildlife Area, which is managed by the Kansas Department of Wildlife, Parks, and Tourism, and an adjoining area now owned by the Nature Conservancy together constitute the largest wetland in the interior United States, one that, incredibly, hosts almost half of all North American shorebirds during their annual migrations. The global significance of Cheyenne Bottoms as a staging area for shorebirds and other waterfowl was recognized in 1988 when it became a Wetland of International Importance under the Ramsar Convention of 1971, a landmark intergovernmental agreement that both signaled the arrival of wetlands as objects of environmental preservation and acknowledged the ways in which migratory birds, as hemispheric actors, demanded global as well as regional and national conservation solutions. In a state where 98 percent of the land is privately owned, much of it an agricultural checkerboard, Cheyenne Bottoms is one of the few large parcels in public land conservation. Saved from agricultural drainage in the late 1920s and early 1930s, Cheyenne Bottoms is an island of nature in a sea of agriculture. Or so it might seem.

I first discovered Cheyenne Bottoms years ago while I was a graduate student at the University of Kansas. A newcomer to the Great Plains, I searched the map for places where I might be able to get out into nature. The pickings seemed slim, but Cheyenne Bottoms was a revelation, the place where I first started paying attention to birds. As I did so, I began to notice that Cheyenne Bottoms was less a remnant of "nature" than a highly engineered environment. Like its surroundings, it was a gridded space with sculpted pools, dikes, dams, renovated islands, and diversion canals that draw water from the Arkansas River to keep the area perennially wet. As a natural depressional wetland, Cheyenne Bottoms might have flooded only one in three years before these improvements, leaving the birds high and dry the other two. In the past, the many other depressional wetlands that once pockmarked central North America made up for its ephemerality, and migratory birds could easily find refuge elsewhere. But as settlers relentlessly drained those areas, Cheyenne Bottoms became more critical as migratory habitat, and managers had to redesign the landscape to serve the birds that consistently concentrated there. These same managers also discovered that the birds were not entirely averse to the surrounding agricultural habitat and its concentrated sources of food, so they began planting crops such as wheat and millet at Cheyenne Bottoms to support the birds, and they began mowing and burning vegetation less favored by the birds to optimize the landscape for their needs. Thus understood, the remarkable concentration of migratory birdlife at Cheyenne Bottoms is not a product of its preserved natural state but of the environmental changes made by the area's managers, many of which were borrowed from the irrigated agriculture that has come to dominate the region. It would be easy to settle on the irony of this situation, that this place of apparent nature is actually a product of human culture, but such irony, and the logic of the nature-culture binary upon which it relies, would miss a crucial part of the story: the roles that the birds themselves—avocets, dowitchers, phalaropes, sandpipers, stilts, and even whooping cranes—have played, through their preferences and behaviors, in the shaping of Cheyenne Bottoms.

If the Weyerhaeuser Environmental Books series has had a central theme and mission during its quarter century in existence, it has been to use historical methods and habits of mind to break down and expose the problems with the nature-culture binary. Perhaps not surprisingly, histories of wetlands—liminal spaces that are neither land nor water—have featured prominently in the series, from Nancy Langston's history of the Malheur National Wildlife Refuge, *Where Land and Water Meet*, and Robert Wilson's historical geography of birds and wetlands along the Pacific Flyway,

Seeking Refuge, to David Biggs's chronicle of the Mekong Delta, *Quagmire*, and Sarah Hamilton's study of Valencia's Albufera National Park, *Cultivating Nature*. Not only have these watery spaces defied and destabilized our reigning environmental categories, but they have taught us to think in critical and creative ways about what it means to "save nature." Nature and culture are so hopelessly entangled in these places, with their histories of diverse human communities in relationship with other species, that the binary itself becomes incoherent.

Emily O'Gorman's *Wetlands in a Dry Land: More-Than-Human Histories of Australia's Murray-Darling Basin* is an essential addition to this list of wetland histories, and to the larger nature-culture conversation, for several important reasons. First, it situates that conversation in Australia, a continent with a distinctive environmental history and rich environmental historiography, but also one that North American audiences have neglected. As evidence, I would point out that among the fifty-five books that now constitute the Weyerhaeuser Environmental Books list, *Wetlands in a Dry Land* is the first to focus on Australia.[1] More specifically, *Wetlands in a Dry Land* introduces readers unfamiliar with the region to the Murray-Darling Basin, the largest on the continent, which drains the arid inland of southeastern Australia. The Murray-Darling Basin is Australia's agricultural heartland, and like the river systems that run through the North American West, it is highly engineered—there are 105 large dams scattered throughout the basin—and its waters are extensively appropriated. But along its length, from source to mouth, it also features an unusual concentration of wetlands, sixteen of which are protected, like Cheyenne Bottoms, by Ramsar designation. These wetlands in a dry land are replete with history and human-animal relations.

O'Gorman takes on a historical narrative that has undergirded wetlands conservation since the term "wetlands" was first coined more than half a century ago. Prior to the mid-twentieth century, expansive settler societies of the sort that came to occupy North America and Australia saw such watery places as obstacles to progress and threats to human health. They were swamps, mires, morasses, unruly and troubled geographies that required reclamation, and these societies became incredibly efficient at ridding the world of such places. As O'Gorman points out, 87 percent of the world's wetlands have disappeared in the last three hundred years, with more than half of those meeting their fates in the last century. But then these very societies came to recognize the value of these watery places—their biodiversity and

FOREWORD | xi

the many ecosystem services that they provide—and set about preserving what was left. As these societies slowly shifted from vilifying to appreciating such landscapes, the word "wetlands" was born as a term for rebranding diverse watery topographies as worthwhile. Thus told, this narrative arc from swamp drainage to wetlands conservation has pitted the nature of wetlands against the human cultural imperatives that have long imperiled them, and it has focused on conservationists and engineers as the primary protagonists and antagonists. It seems a classic tale of the rise of modern environmental appreciation, a heartening story of finally seeing the light.

As O'Gorman traveled through the Murray-Darling Basin and settled into the archives, however, she started to recognize that this story was broken. Not only did the nature-culture binary and its framing of a debate between conservation and development fail to capture the historical complexity of these landscapes; it was, as she notes, "having adverse effects on them." Like Cheyenne Bottoms, most of the wetlands in the Murray-Darling Basin have long histories of human manipulation; the claim that they were or are pristine nature is specious. And one of the most important of those adverse effects was the way in which the binary logic of protecting "nature" from "humans" had marginalized Aboriginal peoples, their material relationships with these wetlands, and their ways of knowing what they call Country. Indeed, this narrative of wetlands appreciation and conservation perpetuated the settler erasure of Aboriginal land uses and land claims that had made these wetlands more-than-natural spaces. As importantly, the binary logic of Western conservation could not accommodate Aboriginal ways of living on Country, and as O'Gorman masterfully shows, it erased other racial and gendered conflicts as well.

If one of O'Gorman's great accomplishments is to complicate the cultural side of the equation, another is how her multispecies approach challenges our understandings of both nature and history. Her impulse to see history as more than human is partly rooted in a foundational claim of environmental history: that humans are not the only actors that matter in explaining change over time. But in making such a case, environmental historians have too often fallen back on a nature-culture logic that we thought we were challenging; by insisting that nature acts in history, we have perpetuated a sense that nature and history remain separable. O'Gorman's more-than-human approach is quietly radical in asking us to collapse these categories, to see all histories as cocreated across species divides, and to recognize that questions of environmental preservation and human justice are inseparable. Her approach is not only increasingly important to the environmental humanities; it is an approach that questions the very concept of the humanities as something we can separate from our studies of the rest of the world. As you will see, O'Gorman has found the "voices" of ducks in the archives, the evolving influence of mosquitoes in human visions of landscape transformation, the rippling effects of nineteenth-century capitalism in the contemporary behavior of seals, and the conjoined salvation of landscapes and cultures in Aboriginal weaving practices.

Why does seeing the world this way matter? For one, it engages with plural and shifting knowledges, including non-Western ways of knowing and being. It thereby points to the need to extricate Western conservation from the jaws of a continuing colonial logic in which it remains trapped. Before reading O'Gorman's book, for instance, I had never thought to ask why Cheyenne Bottoms is called *Cheyenne* Bottoms. O'Gorman helps us see such erasures hiding in plain sight. Another reason this multispecies approach matters is that it better frames the ethical questions raised by these wetlands and their histories, particularly in this new epoch we have called the Anthropocene. The Anthropocene is a powerful frame, and at its best it should shock us into humility, but it can also be totalizing in two troubling ways: it portrays a powerful and disaggregated humankind as a geological force remaking the world, and it makes little room for the persistence of the more-than-human. It is the wetlands appreciation narrative on steroids. Through its elegant series of themed chapters and case studies, *Wetlands in a Dry Land* asks us instead to reimagine just and sustainable futures that are built on interrogating dominant ontologies and honoring Aboriginal ones, and on seeing the histories of all beings as hopelessly entangled. To view history as a continual process of multispecies worldmaking can be disorienting, but O'Gorman helps us see its liberating potential. Hers is an environmental history that sloughs off the nature-culture binary and embraces new ways of thinking about our past, present, and future.

ACKNOWLEDGMENTS

I am deeply grateful to the many people who have shared their knowledge and connections with wetlands in the Murray-Darling Basin with me. Many Aboriginal Elders and custodians have generously spent time with me in conversation and on Country in the course of this research. I would like to especially thank Danielle Carney Flakelar, Meryl and Peter Mansfield-Cameron, Phil Duncan, Bradley Moggridge, and Grant Rigney, along with a number of Aboriginal custodians who wish to remain anonymous. Many other people who live and work closely with wetlands have also spent time with me in conversations and interviews. I would particularly like to thank Wendy Easson, Garry Hera-Singh, Tracy Hill, Matthew Herring, Jamie Pittock, Tracy Rogers, Cath Webb, and Cameron Webb, as well as a number of anonymous participants. Danielle, Meryl, Tracy Hill, Jamie, and Garry also commented on draft chapters.

Other people have given advice and feedback either in conversation or on written chapter drafts. I began planning and writing the first draft chapters for this book during a Writing Fellowship at the Rachel Carson Centre, Ludwig-Maximilians-Universität, Munich in 2014–15. I would like to thank the staff, other Fellows, and Visitors there for their comments, which had an early influence on the overall form of the book. I would particularly like to thank Harriet Ritvo, Etienne Benson, Cindy Ott, Ursula Münster, Celia Lowe, and Veit Braun. Visiting positions in 2018 in the history programs at MIT, facilitated by Harriet Ritvo, and the University of Hawai'i, facilitated by Keiko Matteson, also provided important writing time along with feedback on presentations and informal conversations with scholars there. I have discussed this book at some length with many of my colleagues in Geography and Planning at Macquarie University. I thank them for their support of the project and valuable suggestions. I would particularly like to thank members of the "Environments, Societies, and Power Research Cluster" for their comments on a draft chapter. I am also grateful for the thoughtful comments of Warwick Anderson, Margaret Cook, Heather Goodall, and Grace Karskens on draft writing. Conversations and a co-authored article with Andrea Gaynor have been important in developing my thinking about "more-than-human histories". Many other people have offered useful insights and provided sounding boards for ideas, including Alessandro Antonello, Nancy Cushing, Jodi Frawley, Katie Holmes, Ruth Morgan, Tim Ralph, Libby Robin, and the late Deborah Bird Rose.

Due to the locations of particular wetlands, I have needed to undertake historical research at a number of different institutions. I would like to thank the staff at the State Library of South Australia, Goolwa Historical Society, Alexandrina Library, State Records South Australia, State Library of Queensland, Queensland State Archives, Toowoomba Historical Society, Toowoomba Library, State Library of Victoria, Public Record Office Victoria, Australian Academy of Sciences, Australian Museum, National Library of Australia, and National Archives of Australia. At each of these places, staff knowledge of the collections was important for locating relevant documents. The research for this book was supported by funding from a Macquarie University New Staff Grant and an Australian Research Council Discovery Project (DP160103152).

The University of Washington Press has expertly guided the manuscript into publication. Paul Sutter's advice and comments on the draft manuscript have been invaluable and Andrew Berzanskis has steered the book skillfully through the publication process. Richard Morden created and illustrated the map of the Murray-Darling Basin.

My mother Brit Andresen has discussed many aspects of this book with me, providing thoughtful insights at crucial moments. The person who has had the most conversations with me about this book, and also read several draft chapters, is my husband Thom van Dooren. I am grateful for his support and our many inspiring discussions.

ABBREVIATIONS

CSIR	Council for Scientific and Industrial Research
CSIRO	Commonwealth Scientific and Industrial Research Organisation
DDGGA	*Darling Downs Gazette and General Advertiser*
DDG	*Darling Downs Gazette*
DH	*Daily Herald* (Adelaide)
MIA	Murrumbidgee Irrigation Area
NAA	National Archives of Australia
NPWS	National Parks and Wildlife Service
NSW	New South Wales
PROV	Public Records Office Victoria
QSA	Queensland State Archives
SLQ	State Library of Queensland
SMH	*Sydney Morning Herald*
SOA	South Australian Ornithological Association
SRNSW	State Records of New South Wales
SRSA	State Records South Australia
TCDDGA	*Toowoomba Chronicle and Darling Downs General Advertiser*
WCIC	New South Wales Water Conservation and Irrigation Commission

WETLANDS IN A DRY LAND

INTRODUCTION

MY FIRST EXPERIENCE OF THE COORONG LAGOON WAS FROM THE water. The boat I was in moved downstream across the calm freshwater of Lake Alexandrina toward a barrage that marked our entrance to the Coorong, a Ramsar Wetland of International Importance located at the end of the Murray River system on the coast of South Australia. At the barrage, we entered a lock, which permitted boats to travel between the two water bodies without allowing their waters to mix. The barrages—five in total— were approved by the Murray River Commission (a cross-border, river-focused organization composed of federal officials and state representatives from South Australia, Victoria, and New South Wales [NSW]) and built by the South Australian state government in the 1930s. The barrages were intended to protect agricultural interests upstream by stopping salty water moving from the Southern Ocean, through the mouth of the Murray River, into the northern end of the Coorong, and from there into the lake and, crucially, into the Murray River, along which irrigated agriculture was expanding. The barrages have also prevented freshwater from flowing down the river and into the Coorong, which has since become increasingly saline.

While we waited for the transfer of water to take place in the lock, we watched the long-nosed fur seals lounging along the barrage, some occasionally flopping into the water to catch fish nearby. Their presence heightened our anticipation of the salty waters beyond. A few minutes later we were through the barrage. The water level dropped. We had passed through a threshold into an entirely different place. I could smell and taste salt. As we moved farther into the Coorong, the shores flattened into sand banks where many birds, such as pelicans and sandpipers, waded. The vegetation changed from tall trees and grasses to sedges, rushes, and low, dense shrubs. We passed commercial fishing boats and recreational fishers casting rods from the shore. We turned a bend, narrowly avoiding a sand bank, and there was the mouth of the Murray River. It was open to the ocean. Sometimes, when not enough freshwater is let though the barrages—to preserve it for irrigation use—the mouth closes up. At these times the Coorong becomes even more saline, further and significantly disrupting the lives and reproduction of fish along with other animals and aquatic plants. We continued traveling south along the Coorong, with sand dunes separating us from the ocean on one side and a flat, salt-pan landscape on the other. We stopped along the sand dunes to visit a midden, a large mound of pipi shells made by Ngarrindjeri Aboriginal people over thousands of years. We walked over the dunes to the beaches that stretched along the coast and there dug for pipis with our feet at the waterline. The sun was getting low in the sky. This is where we turned back. Although we had only made it a short way down this narrow, shallow lagoon, it was already clear that this was a place rich not only with wildlife, plants, and striking landscapes but also in histories that profoundly and centrally involved both humans and nonhumans.

Each visit to the Coorong, along with further archival research, revealed new layers to its history and possible future. Over the following years, as I talked to people who lived and worked along the Coorong, I started to form a deeper understanding of the significance of what I had seen on my initial first short trip to this landscape. Ngarrindjeri people and fishers spoke of the impact of the barrages on wildlife, fish, and vegetation. They told me that the seals were a newer disruptive presence, only arriving in large numbers starting in 2007; their prior presence in the area was contested. In the late eighteenth and early nineteenth centuries, sealers had decimated seal populations along the nearby coastline, but in recent decades long-nosed fur seal numbers seemed to be rebounding. These seals were now visiting the Coorong, at least partly because the fish caught in fishers' nets presented a ready food supply. The seals have reduced fish hauls and damaged nets, as well as causing injuries to wildlife such as pelicans. Through further research it became clear that pelicans and other birds were crucial in understanding the history and effects of enclosing parts of the Coorong and its islands as protected areas, including private and government prevention of Ngarrindjeri people's harvests of their eggs. This Ramsar wetland, which we might be tempted to see as natural, is in fact deeply historical, shaped by entangled changes to human and nonhuman lives.

Eight hundred and fifty kilometers upstream, along the Murray and its tributaries, are the checkerboard farms of the Murrumbidgee Irrigation Area (MIA), Coleambally Irrigation Area, and Murray valley irrigation districts. Parts of these areas grow rice by flooding paddies to germinate seed and encourage growth at particular points in the life cycle of these crops. Government water officials divert water that would otherwise flow to local as well as distant wetlands—like the Coorong—to grow rice in the region. During dry times, for example during the Millennium Drought, which lasted from approximately 2000 to 2010 in eastern Australia, government decisions over where to prioritize water delivery—either to wetlands or to farms—have become extremely contentious. Driving around these areas during that drought period, I saw many signs hung by those involved in the rice-growing industry protesting government decisions to reduce water allocations to these farms and instead allow water to flow to rivers and wetlands. One sign read, "You can't eat a wetland." Yet, as many Aboriginal people will tell you, you can eat a wetland by cultivating and harvesting plants in them, hunting animals such as ducks, and catching fish. From the 1970s, the management of wetlands by governments in Australia and elsewhere has focused on their conservation as natural places, largely framing them as separate from humans and culture. This has ultimately served to pit these places against farmers and to undermine understandings of them as valuable to many people for a range of reasons, including denying the long histories of Aboriginal groups' connections to particular wetlands.

Even farther upstream from the Coorong—1,500 kilometers along the Murray, then up the Darling and Macquarie Rivers—are the Macquarie Marshes. On one of my visits to this Ramsar-listed wetland, as I drove along the winding dirt road with a group of environmental scientists, the kangaroos kept pace with our car. Finally on foot, we waded into the reed beds of the marshes—some of the largest in Australia—closely observing the water hens that were moving nimbly along the tops of the reeds. The extent of these reed beds was even more apparent from the air. Joining some scientists on a helicopter flight, I could see reeds out to the horizon. Back on the ground, we listened to Danielle Carney Flakelar, a Wailwan Aboriginal woman whose ancestors had deep connections to this place, a relationship that she and others continue today. She told us about weaving workshops she had held in the marshes to share knowledge within and between diverse Aboriginal cultures, as well as about some of the contemporary challenges to caring for her Country—a term used by Aboriginal people to refer to the nourishing interconnections between multiple beings—including from cotton farms that farmers continue to establish with government approval, both encroaching on the marshes and taking water. Indeed, irrigation, drainage, and other diverting of masses of freshwater by people are crucial for understanding wetland history and futures, especially the way the category of wetlands emerged and has operated in Australia and around the world.

These three sites—the Coorong, the rice-growing areas, and the Macquarie Marshes—are all located within the Murray-Darling Basin, a complex of linked rivers in four states and the Australian Capital Territory in eastern Australia. This massive and highly engineered river system, the largest in Australia, is the major agricultural water source for much of nation and, as a system subject to irregular but major flooding, it has numerous wetlands along the course of its waterways, many of which are of international importance. Most are freshwater, but at the Murray's mouth in South Australia, the ocean waters of the Great Australian Bite meet the river waters, although this meeting is now obstructed by the strictly controlled barrages. Here is the culturally and ecologically important estuary wetland of the Coorong. There have been significant and ongoing losses of wetlands in this river basin since the late nineteenth century, as there have been around the world, due to drainage, the development of irrigation, the construction of associated dams, and the expansion of urban areas. But as my visits to these three sites showed me, old debates about development versus preservation, culture versus nature, do not capture the complex histories and contemporary realities of these areas, and indeed are having adverse effects on them. New ways of understanding, managing, and relating to wetlands in the Murray-Darling Basin—and elsewhere—are urgently needed.

The place to begin is a recognition that wetlands in the basin are rich not only in diverse kinds of plants, animals, and insects but also in human histories. Through my research on wetlands, I have gained a greater appreciation of how humans and nonhumans have together formed these places. *Wetlands in a Dry Land* therefore develops a history that is expansive, examining the ways diverse humans and nonhumans have mutually shaped these places, from the late nineteenth century to the present. It aims to provide new perspectives on wetlands and, in so doing, open up possibilities for new kinds of relationships with and futures for them.

The sites discussed in this book include wetlands recognized as internationally significant under the 1971 Ramsar Convention on Wetlands of International Importance (the Coorong, along with Lake Albert and Lake Alexandrina, South Australia; Macquarie Marshes, NSW; Fivebough and Tuckerbil Swamps, NSW); aquacultural and agricultural landscapes (the Coorong; Fivebough and Tuckerbil Swamps and surrounding irrigation farms; Macquarie Marshes); and wetlands drained by British colonists (Toowoomba swamps, Queensland). All of these sites are connected through the rivers and aquifers of the Murray-Darling Basin. They are also tied to each other and many other places in multiple ways, including through movements of biota and ideas.

The Murray-Darling Basin has been the subject of substantial research and scholarship based in human-focused humanities and social sciences as well as the natural sciences. This book is different in a number of ways. First, it develops the emerging approach of more-than-human histories, which emphasizes that these histories and landscapes have been fundamentally cocreated by varied human and nonhuman agents. Histories that have pitted the "nature" of wetlands against the "culture" of agricultural development have failed to recognize these entangled agencies and have, in turn, reified binaries that have gotten in the way of a more just and sustainable form of environmental management. My approach builds on recent arguments by historians and others that we need to understand wetlands not merely as natural but as socioecological landscapes.[1] Further, this approach is in dialogue with, and draws together, recent themes in environmental history with those in interdisciplinary more-than-human and multispecies scholarship.

Second, this book shows the interconnections among past, current, and indeed future changes in and to wetlands. Each chapter traces more-than-human conflicts and cooperation over time while also demonstrating the continued significance and effects of past events, policies, and interactions on wetlands today. These issues include, for example, the ongoing effects of the constitution of wetlands as an international category of conservation, imbued with specific values, in the 1960s and 1970s within a global recognition of the losses of these places. It further highlights the sustained connections of Aboriginal custodians to particular wetlands as they have sought, and continue to seek, to fulfill their obligations to care for wetland sites. Crucially, tracing these

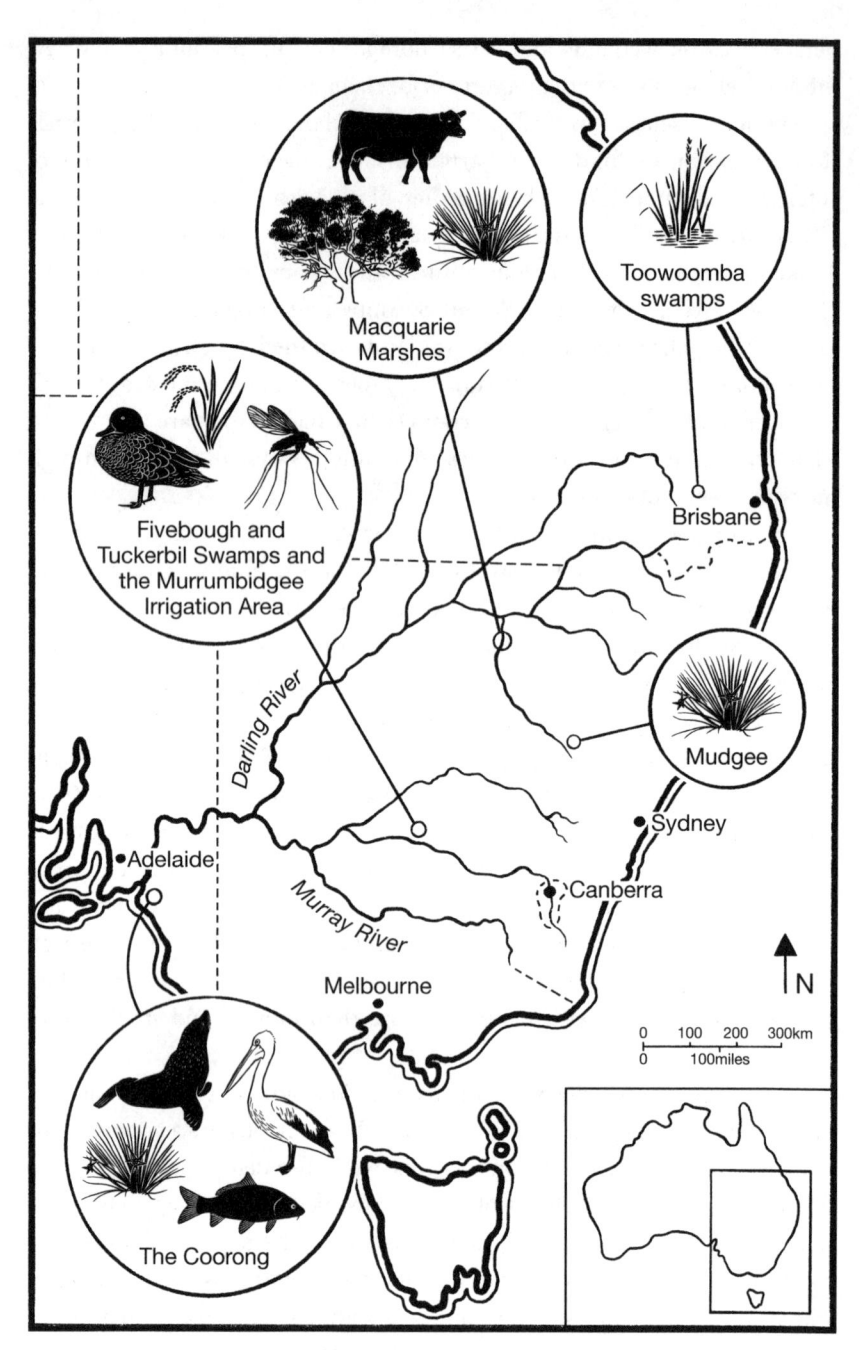

MAP I.1 Map of the Murray-Darling Basin

histories and their legacies shows that the way wetlands have been understood cannot be disentangled from the human and more-than-human lives that have constituted these places. Understandings have been formed through these multispecies relationships and have in turn reshaped them.

A third important difference in my approach is that I foreground the experiences and understandings of Aboriginal people. Key chapters engage with Aboriginal custodians and their histories in relation to specific wetlands, emphasizing the diversity of Aboriginal cultures and their dynamic pasts and presents. These chapters highlight Aboriginal people's experiences, which include actively cocreating these places in relation to and together with an array of plants, animals, and forces. An engagement with Aboriginal histories and practices undermines simple environmental narratives of wetlands as degraded by human activity and rather demands that we attend to the diverse interactions people can have with these places. The chapters engage with the ways Aboriginal people and groups have grappled with the dramatic and continuing changes brought about by colonialism and, more recently, the fluctuations of globally circulating technologies and economic impacts. Indeed, far from distracting from issues and ideas of race, a more-than-human perspective provides a further means for interrogating them, revealing the ways they have been reinforced and resisted in complex human and more-than-human relationships.

Fourth, my approach foregrounds class and gender. Wetlands have been places in which these issues—which intersect with racial ideologies and different views about the relationship between people and nature—have played out and taken shape, in the process remaking these places. Indeed, ideas of class and gender have shaped the diverse values people associate with wetlands as well as changes to these over time. Attention to these issues further reveal the continuing pressures and changes experienced by all populations, colonized and colonizing as well as newly immigrant.

Fifth, chapters examine the roles and practices of various sciences in shaping wider understandings of wetlands. From entomologists studying mosquitoes as vectors of disease to ecologists working toward wetland conservation, scientists have provided particular sets of expertise and kinds of knowledge through which wetlands have been appreciated. Crucially, the chapters show that the focuses of particular sciences have reflected and reinforced specific sets of values at different times.

Sixth, nature—in many ways the central element each of these stories—is treated neither as static nor pristine. Species, as well as individuals within species, react differently to different pressures. Hungry seals, migrating birds, and newly arrived European carp have all lived in wetlands, but their numbers, behaviors, and interest in these places have changed over time as wetlands themselves have changed. Their dynamism as species and individuals counters narrow notions of nature as simply a background to human concerns. As these animals and plants have gone about their lives, they have both altered and been shaped by diverse human projects. Their specific behaviors and needs have helped shape these places and histories. *Wetlands in a Dry Land* aims to foreground the agency of the more-than-human world, inviting the reader to consider how attention to diverse and specific agencies suggests how we might engage differently with wetlands.

Indeed, wetlands have been shaped both with and against other species, from saving birds, to weaving sedges, to eradicating mosquitoes. The diverse values people have associated with wetlands have been formed through their relationships with particular, historically dynamic nonhuman species. Specific and changing understandings of these places—from Indigenous cultural landscapes to gardens, miasmic wastelands, and precious wetland ecosystems—have emerged out of multispecies interactions. The unruly movements of ducks and other waterbirds have impacted farmers. The behaviors of seals have created significant problems for the fishing industry in the Coorong. Mosquitoes buzzing and biting their way through the landscape have frequently defied efforts to fully understand, let alone control, them. Countless other animal and plant lives, along with microbes and others, go on playing their own significant roles in the histories of these wetlands; in all their diverse forms, wetlands are multispecies achievements. Imagining, inventing, and ultimately remaking wetlands has been and remains a multispecies project.

Finally, while this book offers an in-depth history of an important part of Australia, the questions it raises and the conflicts it deals with are global. This book is of direct relevance for all former settler colonies, including the United States, Canada, and New Zealand, but also more widely, including the formerly colonized worlds of India and Latin America and among European former colonizers. Wherever human and more-than-human populations and ecologies grapple with massive economic change, demographic and technological change, and—overarching all—climate change, the issues raised by these chapters will be relevant. This book details clashes between the knowledge and practices of Indigenous peoples and the new economies with which they are dealing. It addresses the class and gendered impacts of changing technologies across all human populations at the same time as these populations interact with nonhumans, including resurgent and declining "native species" as well as "invasive species" (for better or worse). These are all issues faced globally, on larger or smaller scales, even though the local details may differ.

This is a global story, but it takes particular forms in Australia, the driest inhabited continent on Earth. Many regions experience long dry periods punctuated with intermittent wet years in which large floods flow through and temporarily expand watercourses, increasing the number and size of wetlands. Australian environmental and more-than-human histories offer important insights into how people, animals, and plants have lived in and together helped shape such landscapes. Indeed, Australia offers crucial perspectives on ways of living with uncertain water and climate regimes in which dynamic human and nonhuman agencies respond to and thus shape emergent worlds. These perspectives are even more important in this era characterized by climate change and ecological crisis.[2]

Indeed, in these circumstances, wetlands offer vital perspectives. Today ecologists estimate that as much as 87 percent of the world's wetlands have disappeared in the past three hundred years, and over 50 percent of those have been lost since 1900, largely due to drainage to prevent disease and encourage urban growth, along with water diversion to support intensive agriculture.[3] The continued reduction of wetlands globally has had cascading consequences, from biodiversity loss to the erosion of cultural sites. In this context, many wetlands have become sources of contestation—and at times conflict—between various groups and uses: irrigation farmers, Indigenous communities, conservationists, developers, and government officials have all used and valued wetlands in different ways at various moments in time.[4]

These issues are being exacerbated due to changes to wetlands—and shifting understandings of their value and vulnerability—under anthropogenic climate change. Wetland ecosystems, which support approximately 125,500 freshwater species globally, are some of the most biodiverse in the world but also some of the most sensitive to changes in water availability, new species, floods, and droughts. Coastal wetlands play an important role in buffering storm surges; without them the effects of cyclones and hurricanes are much worse for humans, plants, and animals. They are carbon sinks, and being drained, having water diverted, or otherwise being lost releases carbon dioxide as well as methane into the atmosphere, contributing to climate change. Some researchers have even argued for wetlands restoration as a means to store carbon dioxide and lessen the effects of storm surges, floods, and droughts on people and other species.[5] In this and other ways, wetlands draw us into histories and futures of the Anthropocene, a proposed geological epoch in which humans play an increasingly significant role in shaping the planet's ecological, hydrological, climatic, and other systems.[6]

Mirroring and contributing to this global situation, many wetlands in the Murray-Darling Basin have disappeared, or significantly changed in character, due to the intensification of land and water use in the region starting in the late nineteenth century and accelerating in the twentieth century.[7] However, ecologists cannot currently give any precise assessment of the losses of these places. This is partly due to the difficulties posed by the category of wetlands in this region, which has complex boom-and-bust hydrologies influenced by intermittent wet and dry periods that can last years and radically fluctuating numbers of nonhuman species, such as birds, frogs, and fish. Yet ecologists' evidence shows that the basin has sustained a disproportionately high rate of wetland losses in Australia.[8] Climate change further exacerbates the effects of water diversion from wetlands in the basin, as both droughts and floods continue to increase in intensity in this region.[9] In this context of global wetland losses and valuable Australian perspectives, the Murray-Darling Basin and its wetlands—both diverse in kind and large in number—offer unique insights and represent an important global case study.

WETLANDS IN A DRY LAND

The hydrology of the rivers and wetlands in the Murray-Darling Basin is central to their ecology. All the rivers are characterized by variable flows and experience periods of intense drying as well as large floods. Because of this, many wetlands are ephemeral or increase significantly in area during flood flows. During floods, biodiversity in wetlands, along rivers, and on floodplains increases as floods trigger breeding cues for many plants and animals.[10] In the basin, wetlands often see bursts in life during wet years, and some also become important sites of sustenance during dry years—which are the norm—as they continue to hold some water and support life long after other water sources have dried up. In fact, the basin contains over thirty thousand wetlands, with many recognized as being of national importance and sixteen of international importance under the 1971 Ramsar Convention (out of a national total of sixty-five).[11]

The basin is also Australia's agricultural heartland and is often referred to as the nation's food bowl. According to available statistics, in 2005 the region accounted for 39 percent of Australia's agricultural value and contained 65 percent of the irrigated land, with 84 percent of land in the basin owned by agricultural businesses.[12] Today, the basin poses environmental management challenges at a range of scales. Most prominently, these challenges center on competing views over agricultural water diversion from rivers and wetlands for farming, the channelization of rivers, and the construction of irrigation infrastructures, as well as Aboriginal peoples' access to water for a range of reasons, including spiritual, cultural, and economic values and activities. These contemporary debates and physical landscapes have been shaped by past approaches to wetlands, including local uses—such as hunting, fishing, and farming—that have both created bonds with particular wetlands and altered them.

Wetland environments in the Murray-Darling Basin and in Australia more generally need to be understood in the context of Aboriginal people's millennia-long history of living in and with these places. Aboriginal people have cocreated diverse landscapes across the continent over tens of thousands of years, including many wetlands, for example through burning and other agricultural, aquacultural, and spiritual practices.[13] Paying attention to this cocreation disabuses us of notions that these wetlands are simply natural and thus by definition threatened by human activity.

British colonization beginning in the late eighteenth century radically altered Aboriginal people's relationships with wetlands in this region. Despite the often-deliberate efforts by British colonists to sever Aboriginal people's connections with wetlands, many Aboriginal people and groups maintained their relationships with these places. Histories of Aboriginal people's struggles and resistances as well as of colonization have influenced recent postcolonial and decolonizing changes in water and wetlands management in Australia, as recognition of Aboriginal people's knowledge of and rights to lands and waters has grown.[14]

British colonists moved inland from the east coast starting in the mid-nineteenth century, toward what they later called the Murray and Darling river systems, driven by a desire to find new grazing pasture for sheep and cattle.[15] The colonists were drawn to wetlands as important sources of freshwater in this often dry region. While European colonists were thus brought into close proximity to wetlands—which they rather called swamps, marshes, and lagoons—they and others frequently viewed them as dangerous areas, wastelands, and sources of disease. They channelized and later drained many wetlands, modifying water flows and bringing a host of new organisms to these places, with important consequences. In addition, waves of other new migrants, such as those from China during the mid- and late nineteenth century, also altered wetlands, including through establishing irrigated market gardens along their edges and by sinking wells. However, the new kinds of expertise and knowledges introduced by the British (which drew on diverse knowledges from many other places) came to dominate the shaping of wetlands in this region.[16]

Perhaps most significantly, these approaches informed the construction of large dams and extensive irrigation networks by state, interstate, and federal government bureaucracies, particularly across the twentieth century. These were built to support river navigation and, increasingly, agriculture. Together with the long-term pumping of aquifers, these projects have reduced the water available to wetlands. While rivers had been dammed and diverted for agricultural production before this, the rapidity and scale of agricultural expansion and the associated water diversion in the twentieth century were new. In that century many of the 105 large dams that now stand in the Murray-Darling Basin were built.[17] In the Murray River system alone, the Lake Victoria Reservoir, Hume Dam, Snowy Mountains Hydro-Electric Scheme, five barrages on Lake Alexandrina blocking the Murray mouth, thirteen locks spanning the Murray, Lake Mulwala, and other works were completed between 1915 and 1974, all at least under construction by the mid-1950s. While such projects were undertaken across Australia, the Murray-Darling Basin was a particular focus for state and federal governments which aimed to overcome the vagaries of drought, increase agricultural production, and increase settlement in the inland, and which also competed with each other over use of the water from this cross-state river system.[18] Indeed, historical geographers Trevor Langford-Smith and John Rutherford noted in the mid-1960s that the plains of the Murrumbidgee River in NSW, which flows into the Murray River, had "by far the greatest concentration of irrigation development" in Australia.[19] In these changing waterscapes of the Murray-Darling Basin important new kinds of wetland environments emerged with complex histories and ecologies of their own: places like rice paddies and sewage treatment works.[20]

In the 1960s and 1970s, many people became concerned over the decline in wetlands in Australia and globally. Local movements but also, crucially, international efforts to save wetlands were reactions to accelerating losses of these landscapes around the world. Indeed, it was in this period that "wetlands" became an international category and object of conservation amid these campaigns, which reinforced and were supported by global efforts in broader environmental protection. This new global nomenclature was supported by governments and scientists around the world to shed the negative associations of terms like "swamp." They instead deliberately reframed these as precious places that needed to be set aside for conservation. In many ways this represented a radical new mode of imagining these places that positioned them as internationally significant wildlife habitat, especially for migratory waterbirds. Indeed, these global efforts led to the establishment of the Ramsar Convention on Wetlands of International Importance, Especially as Waterfowl Habitat, in 1971. This convention in turn intersected with the reframing and recategorization of these places by scientists and governments in Australia as sites for the conservation of nature. In Australia, an agreement with Japan on the protection of birds that migrated between the two countries further reinforced the government's focus on conserving wetlands as bird habitat. In all of these ways, wetlands in the Murray-Darling Basin have been deeply shaped by past, including recently past, water and environmental politics and cultures.

Most histories related to water in the basin have focused on the lives of water engineers, the construction of water storage and diversion projects, and the role of water bureaucracies in manipulating water flow.[21] These histories have been important in revealing some of the rationales behind and consequences of these schemes. Yet there were many other actors, including a diverse array of humans and other species, that shaped these histories. This book expands traditional narratives by highlighting the diversity of human and nonhuman agencies that have cocreated wetlands. Examining nonhuman agencies further underscores the need for new narratives of colonization in this region, showing how relationships among Aboriginal people, European settlers, and others were dynamically embedded within and shaped by wider more-than-human relationships. In engaging with these approaches and arguments, this book aims to further develop emerging approaches to more-than-human histories.

MORE-THAN-HUMAN HISTORIES

In 2010, I went into the archives looking for government documents on water use in commercial rice growing in NSW and Victoria, but I came out thinking about wetlands. In the context of an ongoing drought, I wanted to know more about the changing water politics of irrigation and rice. How had this water-intensive crop come to be grown and then to dominate agriculture in areas that had previously experienced long droughts?[22] As expected, I found large files on water and rice growing created by the state governments. What I had not expected were the almost equally large files on ducks in rice-growing areas. I knew the annual duck migrations across the rice-growing areas of California had proved challenging for managers there, but the Australian ducks were not migratory.[23] They were instead nomadic, following less predictable movements of water. In the MIA, these ducks presented some unique challenges for managers, farmers, and scientists. For many farmers, various species of ducks had become pests soon after they started growing rice there in the 1920s. These birds damaged rice crops when they settled in paddies and ate freshly sown rice. But for the ducks, the flooded rice paddies were additional watery, grassy places where they could swim and forage—sometimes replacing other wetland habitat as farmers diverted water for this farming. In this way the flooded rice paddies became a kind of wetland, with not only ducks and rice but many other plants, frogs, birds, and insects moving into them. Some farmers welcomed some of these new arrivals, others not. Much depended on the farmer. As I entered farther into the ducks' stories, they offered a different view of the history of rice growing, water, and wetlands, one that gave insight into their needs and values as well as their influence on a range of human concerns. Further, their movement and attraction to water seemed to contest the human boundaries between agricultural lands and natural areas. The question the ducks prompted me to ask is this: What has counted as a wetland, for whom, and with what consequences? For the ducks it mattered that many farmers did not consider the paddies to be wetlands. Special open hunting seasons were regularly declared by the NSW state government on ducks in these areas, as they still are. Ultimately, the ducks started to change my sense of who the important actors were in this history and just how much control humans had over the histories they tried to make.

More-than-human approaches help us understand these sorts of socioecological relationships. These approaches have developed over the last two decades through a transdisciplinary dialogue among science and technology studies, anthropology, geography, philosophy, and other fields. This dialogue has shaped an interdisciplinary set of concerns that have come to sit within the environmental humanities. In general terms, this scholarship rejects notions of human exceptionalism that, in Western traditions, have helped separate "nature" from "culture" and justify human exploitation of an externalized nature. Instead, it argues for richer examinations of socioecological relationships that more fully account for a diversity of actors, both human and nonhuman. Crucially, these approaches emphasize the importance of situating shifting modes of understanding that have been used to make sense of and inhabit these relationships. These modes of understanding, including those in the sciences, have been formed within specific histories, cultures, and more-than-human relationships, and have had particular consequences.[24]

In many ways, environmental historians have shared these concerns. Environmental history is an established field that emerged in the 1970s and has since sought to interrogate changing human relationships with the environment. Indeed, historians have sought to reveal the dynamic roles of other organisms, from Alfred Crosby's "ecological imperialism" to Harriet Ritvo's work on animals in Victorian England, and the contributors to a recent collection titled *The Historical Animal* who argue for the importance of taking animal perspectives and behaviors seriously, including in the making of films.[25] However, these works have largely focused on animal histories, not always explicitly aiming to contribute to multispecies and more-than-human scholarship.

While environmental historians increasingly identify with the environmental humanities, the field has largely treated this as an umbrella rather than an interdisciplinary space that it too can transform and also be transformed by.[26] There have been ongoing calls by some environmental historians for the field to engage more fully with key work—such as by Donna Haraway and Bruno Latour—that has informed more-than-human and multispecies approaches in order to enrich their narratives with a more diverse array of actors (human and not) and more thoroughly situate knowledges.[27] However, practitioners have been reluctant to do so. Indeed, more-than-human scholarship has also been slow to engage with approaches in environmental history. This lack of sustained and explicit dialogue has meant that environmental historians have often overlooked key aspects of multispecies and more-than-human scholarship that complicate their approaches. These approaches often either maintain a division between nature and culture and a focus on human actions, or they do the opposite, failing to sufficiently entangle natural agencies with the human. Indeed, even the hybrid approaches that have come to characterize environmental history still maintain the categories of and separation between nature and culture while seeking to examine their interactions.[28] This means that more-than-human agencies and their historical consequences are not fully examined. Further, as environmental historian Gregg Mitman has noted, though environmental history is committed to relationality, it has been "deeply resistant to embracing a relational ontology in which things exist not in themselves" but rather in "changing material, social, and symbolic relations between and among human and non-human actors."[29]

There is, however, a growing body of historical scholarship that explicitly engages with these multispecies and more-than-human approaches to examine the agency of other biota in shaping (human) histories, while also being attentive to shifting and diverse ways of knowing, human and beyond.[30] These more-than-human histories show that both environmental history and more-than-human approaches are needed to fully account for the diversity of humans and nonhumans that have shaped socioecological relationships and how these relationships have shifted over time. Environmental historians pay close attention to the ways specific relationships—that include humans and nonhumans—have been coconstituted over time, which can help us understand why and how particular worlds were produced out of these relationships. Indeed, to fully appreciate coconstitution, we need histories, as relationships that make worlds are formed and re-formed over time. At the same time, environmental history can learn from interdisciplinary more-than-human approaches, particularly in studying and situating shifting and diverse ontologies and ways of understanding that have helped shape particular worlds within dynamic more-than-human relationships. This book builds on and engages with these emerging approaches and in so doing aims to bring environmental history and the environmental humanities into closer, productive, explicit dialogue.

These more-than-human histories are always situated and partial, and they necessarily run counter to overarching grand narratives. In engaging with particular sets of more-than-human relationships and situated knowledges, such approaches emphasize diversity and multiplicity.[31] In this way, they work to avoid homogenizing and universalizing the human, so as to pay close attention to differences in human experiences, even as they aim to rethink and reposition human lives in the context of multispecies entanglements. Questions of race and gender are deeply entangled in multispecies worlds, which can be meaningfully examined through detailed historical studies.[32] More-than-human histories are necessarily engaged in complex ethical considerations, seeking to find careful paths through and possible ways forward within these dynamic relationships. Rather than arguing for simply protecting nature or defending human activities, they must contend with more mixed consequences. Within this dynamism, they must carefully, and ultimately, argue for particular kinds of understandings and more-than-human worlds over others.

By approaching the past in such a relational way, this book positions humans within temporally as well as spatially dynamic interspecies relationships.[33] It aims to show that being attentive to these relationships—and how they have been imagined and lived by an array of human and nonhumans—can reveal a range of inequities and draw attention to both human and nonhuman diversity. A more-than-human and multiscale history can further reveal the deep roots of inequities and power relationships, highlighting how the ways in which we humans know and act are shaped by inheritances we don't always recognize. Attending to multiple ways of interacting with and understanding wetlands opens up not only new histories but also futures for these places—histories and futures that might better account for and even promote this diversity.[34]

Wetlands are liminal places. They are where land and water meet. But where land ends and water begins is muddy and unclear. Different kinds of waters, from fresh to brackish to salt, can meet and mingle in wetlands, as in the Coorong lagoon. In this way, wetlands prompt a range of questions about borders, crossings, and leakages for various agents, human and not, in various times and places. These issues of borders and transgressions return throughout the book as it examines who has a stake in establishing borders and some of the consequences of doing so, as well as the way wetlands, different species, and people have disrupted these boundaries. This understanding of wetlands resonates with feminist, more-than-human, and multispecies ideas about bodies not ending at the skin but rather being "leaky" and fluid.[35] Much is at stake in these borderlands and their crossings. This book seeks to better understand these places and their boundaries by engaging with diverse human and more-than-human relationships, examining the ways they have shifted and endured over time. We might regard wetlands, then, not so much as separate places but as far-reaching and temporally rich entangled multispecies processes.

STRUCTURE OF THE BOOK

Each chapter of *Wetlands in a Dry Land* centers on a key theme that illuminates an important aspect of the history of wetlands in the Murray-Darling Basin and beyond: weaving, leaking, infecting, crossing, enclosing, migrating, and rippling. These are themes that highlight relationships, especially the socioecological entanglements of humans and nonhumans, many of which take place across and complicate Western modernist human borders. These accounts draw on original archival research as well as interviews and time spent in these landscapes. In each case, I have sought to read and listen for multiple agencies. While many of the chapters focus on specific animals and plants, these discussions are situated within a bigger cast of organisms and forces to show the multiple more-than-human relationships at stake.

The chapters' thematic focus allows each to bring together past and more recent events, policies, and interactions. The book begins with contemporary concerns, in order to foreground the pressing issues that Aboriginal people, particularly women, grapple with as they seek to create healthier Country amid multiple pressures and historical legacies. By focusing on *weaving*, chapter 1 shows that the concerns of Wailwan and other Aboriginal groups—concerns that may seem disparate to others—are in fact intimately connected. It aims to provide a complex, richly woven understanding of what is at stake in the lack of water in wetlands. Weaving interconnects generations and Country, water politics, and access to wetlands.

The remaining chapters are roughly chronological, focusing on successive events and particular wetlands that are important in understanding the history of wetlands in this region and elsewhere from the mid-nineteenth century. In chapter 2, I focus on *leaking* in the city of Toowoomba, in southeastern Queensland. Toowoomba is a town that was built on swamps. This chapter considers the swamps as historical actors, focusing on disease as one of the key ways they shaped this township from 1850 to 1940. At the same time, it takes up important questions of race and class, examining interwoven processes of swamp improvement and control, many of which ultimately failed, creating new health problems for humans.

Chapter 3 centers on *infecting*, analyzing changing understandings of mosquitoes and irrigation systems in light of a series of government investigations into the possibility of an outbreak of malaria in the MIA between 1919 and 1945. Engaging with the context of two world wars and increasingly militarized approaches to environments, this chapter focuses on how the mosquitoes reveal and connect the shifting biocultural terrains of agriculture and nature, the home and nonhumans, as well as approaches to race and health.

In chapter 4 I look more closely at the *crossing* of borders by wildlife into agriculture. This chapter focuses on the MIA's changing relationship with Fivebough and Tuckerbil Swamps, which are nestled within its expansive rice farms, established there starting in the 1920s. It examines the highly consequential work of establishing and maintaining the boundaries between agriculture and wildlife, with a particular focus on the repercussions of this boundary work for the native species of ducks that have sought to move within and across these landscapes.

Chapter 5 focuses on the *enclosing* of wetlands, a practice that has a long history in many places and significant implications for Aboriginal people in Australia. This chapter centers on the fallout from a mass slaughter of pelicans in 1911 on a group of islands used as rookeries in the Coorong lagoon, South Australia, highlighting the way ideas of race and private property have together played a role in establishing protected areas. In so doing, the chapter foregrounds the complex interplay between the values and approaches of conservationists and Aboriginal people, an interplay that has frequently shaped wetland environments and possibilities in Australia.

Chapter 6 examines the role of *migrating* birds in shaping the emergence of wetlands as a category of conservation in the 1970s. This historical situation has had lasting consequences. After joining the Ramsar Convention on Wetlands of International Importance and the Japan-Australia Migratory Bird Agreement in this period, Australian government conservation departments initiated a nationwide Wetlands Survey. This survey was ultimately abandoned, partly due to disagreements over whether wetlands were primarily an object of conservation as waterbird habitat or to be used for other uses and values.

Chapter 7 engages with sealing in the late eighteenth and early nineteenth centuries as rippling capitalism, with ongoing effects, importantly shaping contemporary controversies in the Coorong lagoon area that in many ways are hallmarks of the Anthropocene. In this region, long-nosed fur seals have become the center of a heated debate as their numbers exploded over the last decade, causing financial and cultural upheaval for Ngarrindjeri people and fishers. Some fishers have turned to invasive European carp to recoup their economic losses. Is it possible or even desirable to manage wetlands toward their state prior to British colonization?

The afterword brings key issues and themes raised in the chapters into dialogue with the long dry period experienced by humans and nonhumans throughout the basin culminating in the summer of 2019–20, and into a consideration of future changes in these wetlands in the context of climate change. In undertaking a multiscale and more-than-human history of wetlands, this book tries to widen the scope for reimaging these places as they continue to be lost, to show that things might have been, and might still be, different.

ONE

WEAVING

Postcolonial and Multispecies Politics of Plants

I WALKED TOWARD A STAND OF REEDS ON THE EDGE OF A CATTLE-grazing property. It was July 2017, and I was on the Country of Wailwan (also spelled Wayilwan or Weilwan) Aboriginal people. The reeds were part of an extensive complex of rivers, reed beds, and floodplain river red gum forests that made up the Macquarie Marshes, located in north-central NSW. The Macquarie Marshes are a Ramsar-listed wetland, recognized internationally for their extensive reed beds and ability to support large populations of diverse plants and animals. The ecology of the marshes has been partly shaped by a long history of intermittent periods of drying, which can last years, as well as flooding, which can last months. Many of the plants and animals are highly responsive to these changes; for example, many waterbirds begin breeding en masse in the marshes when they are in flood.[1] The marshes are important in Wailwan culture for many reasons, including Wailwan women's use of semiaquatic reeds and sedges for weaving. They are bordered in many places by grazing properties and in other parts by water-intensive cotton farms.

As I made my way toward the reeds, I wondered, Would there be water in the reed beds or would they be dry? As I followed muddy cattle tracks into the reed beds, small pools of water became deeper lagoons and rivulets,

FIG 1.1 The Macquarie Marshes are bordered in many places by cattle grazing properties. Cattle grazing and the establishment of these properties have historically influenced Wailwan people's access in complex ways. Cattle have also affected the ecology of the marshes. Photo: Emily O'Gorman, 2017.

lined by clumps of reeds that provided the only trusted footholds. I was there with a group of mostly Australian and South African environmental scientists, and we all delighted in the muddy wateriness. It was a relief after hearing news stories earlier that week exposing corruption in the state government, which allegedly had facilitated illegal water siphoning by irrigated cotton farmers in this area, part of the wider water politics of the Murray-Darling Basin.[2]

We traveled into other parts of the marshes and their floodplain with Danielle Carney Flakelar, a Wailwan woman whose father's family has lived, worked, cared for, and managed this Country for a thousand generations.[3] Danielle is Wakka Wakka on her mother's side. She grew up with her father's family in Wailwan Country. Danielle told us about the long and dynamic Aboriginal history of the area, in part revealed though a number of distinctive mound sites—low rises in the flat floodplains—which, she told us, were created by Wailwan people through ongoing use, being built up over time through everyday acts of living, telling stories, and performing ceremonies. Aboriginal people arrived in northern Australia many generations ago, and Wailwan people have lived in the Macquarie Marshes area for tens of thousands of years.[4] Over this period, they have fundamentally shaped the shifting ecology of the marshes in multiple ways, including burning the landscape and cultivating and harvesting plants.[5] The mounds we visited were close to the "Old" Macquarie River, now mostly dry; the river had gradually made a new course as the water took a different path across the floodplain, and most of the remaining flows had been diverted for irrigation.[6] Being in the marshes listening to Danielle, with the environmental scientists, as local, basin-wide, and national water politics unfolded, I started to think about the marshes from multiple perspectives: multiple species and beings, multiple lives and generations, multiple temporalities and scales, multiple values and relationships that all interwove to shape this place where we were. The Wailwan dreaming story that describes the creation of the marshes is, in fact, about relationships among different beings and their effects on one another, in which generosity and greed are central themes.[7]

When I next spoke to Danielle, just over a year later in August 2018, she said, "I can see the marshes dying in front of me."[8] The state government had declared most of NSW to be in drought, and the marshes were drying out quickly. Droughts in eastern Australia have become longer and drier under anthropogenic climate change. But there were other, related factors at play here as well. The management of the Macquarie Marshes and the Macquarie River, which flows to and through them, has been an ongoing source of contention among pastoralists, irrigators, Wailwan people, and conservation groups in the area for more than one hundred years. Danielle noted that Wailwan people were silenced and treated as if they did not exist and therefore were not included in conversations about Macquarie Marshes water management regimes until the late 1970s and early 1980s, and they still fight for sustained, meaningful input into these decisions. Contestations over water management in the area gathered pace from the 1960s, when water-intensive cotton farms started to be established nearby and the long-debated Burrendong Dam was completed upstream of the marshes in 1967.[9] Together, these have had ramifications for a range of animals, plants, and people. Declines in waterbird numbers, reed bed area, and river red gum health, along with the availability of sedges and reeds due to reduced water flows, have had cascading effects on various beings and cultures.[10]

As we continued our conversation, Danielle said, "They're still clearing out there now—land that has never been cleared—and they're putting ... cotton in. I've had to find a way to regather my strength to be present in Country." Aboriginal people's concept of Country has sometimes been described as traditional lands and waters but it is much more than this, encompassing specific relationships among multiple species and beings in mutual life-giving.[11] Danielle was making plans for a group of Aboriginal women to go to the marshes later that year to share their knowledge about managing Country and culture through medicine, language, and weaving. But, she said, "I'm not sure what state the marshes are going to be in or where we're going to have to go to access a little bit of sedge [for weaving] and how much will be available."[12] At a conference the previous year, I had heard Danielle speak about the importance of negotiating with the NSW state government for water flows to be released from Burrendong Dam to the marshes to encourage the growth of wetland plants used in weaving, such as spiny sedge, and trees, such as river red gums, which rely on intermittent periods of flooding and drying. Danielle spoke about the importance of the timing of this water in relation to organizing Wailwan burning of particular areas, which further encouraged reeds, rushes, and sedges to grow new, strong fibers. Most of these plants, like spiny sedge and cumbungi rushes, grow near or in water. These plants provide food and habitat for a range of animals, including certain birds and frogs, as well as other plants. They filter water and stabilize banks, and in taking care of them, others are provided for too.[13] In turn, some of the plants provide fibers for Wailwan weaving. Weaving is an important aspect of Wailwan knowledge. It can also lead to wide-ranging discussions during group harvests and shared practice around spirituality, place, and responsibility. In these and many other ways, weaving is part of being on and caring for Country.

To stay with these important stories of Aboriginal women that are unfolding within and continually remaking wetlands Country, this chapter has a more contemporary focus than other chapters in this book. These contemporary stories urgently need to be heard because they bring to light the need for important changes in water management regimes and greater recognition of Aboriginal women and people, along with their knowledges, in management regimes and legislation to enable healthier Country. At the same time, these are inescapably historical stories. They are at once pressing and multitemporal, multigenerational. They hold manifold temporalities that are human and more-than-human.[14] They are deep rooted. By focusing on weaving, this chapter shows that the concerns of Wailwan and other Aboriginal groups, which may seem disparate to others, are in fact intimately connected. In this way it aims to provide a complex, richly woven understanding of what is at stake in the lack of water in wetlands.

Wailwan people and many other Aboriginal groups have long, ongoing connections with wetlands in the Murray-Darling Basin.[15] These places have been and remain important sites of political engagement and activism for Aboriginal people as they continue to seek to fulfill obligations to Country, including strengthening their rights and roles in dominant Western water and site management processes and practices. This chapter engages with the contemporary activities of Aboriginal women at three wetland sites in the Murray-Darling Basin—the Macquarie Marshes, the Mudgee area in NSW, and the Coorong in South Australia—as they use weaving of sedges and rushes to show the importance of these places to their kin, community, and other groups, such as governments, and to continue connections to Country that have persisted through British colonization and up to now. It threads together the stories, knowledge, and experiences they have shared with me into one of many possible forms. Parts of my own story get entangled and woven in here too, including a range of strands that have shaped my thinking about more-than-human histories.

Plants like sedges and rushes hold a postcolonial and multispecies politics. They play an important role in, and provide a lens into, the historical and ongoing connections of Aboriginal women with particular places, cocreating and interconnecting worlds.[16] The first part of this chapter examines the way weaving involves multiple generations and Country. The last half of the chapter then specifically focuses on the way weaving has been, and continues to be, interconnected with water politics and access to Country in the Macquarie Marshes area. It does this in conversation with Danielle, who has shared important insights with me about these issues.

Weaving is never simply the making of an artifact. Rather, it is a temporally dynamic multispecies practice of world-making, through which particular more-than-human histories and futures take shape and are expressed. Anthropologist Tim Ingold has shown that woven coiled baskets complicate the neat divisions made in many academic disciplines between materials and culture. Instead, he shows that "the surface of nature is . . . an illusion," because we all work "from within the world, not upon it." For Ingold, "Weaving focuses on the character of the process by which that [woven] object comes into existence," shaped through a range of pressures and resistances that are both human and nonhuman.[17] Plants, along with people, other species, and forces, shape the weave. Sometimes weaving does not include humans at all, as demonstrated by the efforts of male weaver birds in parts of Africa who use leaf fibers, twigs, and grasses to form elaborate nests—or, we might add, male and female willie wagtail birds in Australia, who weave nests from grasses, spiderwebs, and animal hair or fur. Ingold argues that weaver birds show judgment and care as well as intergenerational teaching, which draws us into a consideration of other species' cultures, also formed within distinct multispecies relationships.[18]

This chapter centers on a limited number of places and in conversation with particular people. However, throughout the Murray-Darling Basin and in many other places across Australia, Aboriginal women have ongoing weaving practices and connections to Country.[19] As Kate, a Wiradjuri woman from Mudgee, NSW, said to me, "There are lots of communities doing things [with weaving and revegetation of weaving plants], and if not there are individuals doing it."[20] In fact, a relatively recent revival in weaving practices has meant that spiny sedge is now being grown by Aboriginal women in places on the South Coast of NSW, where it was not evident prior to British colonization in the late eighteenth and early nineteenth centuries.[21] This is part of a long history—extending across tens of thousands of years—of Aboriginal people shaping the vegetation of large parts of Australia through burning, aquaculture, and agriculture, including practices related to or affecting weaving plants.[22] Indeed, it is important to recognize that these practices, more than just meeting domestic needs, constitute environmental stewardship and care for Country.[23] Care for Country encompasses an ethics of spiritual care and mutual nourishment between beings. These understandings have been and are world-making, cocreating entire landscapes in dynamic more-than-human relationships.

As an engagement with weaving demonstrates, Aboriginal women are leading creative solutions in wetlands management and politics. They have often taken on important roles in these spheres, as well as wider responsibilities in caring for Country. However, the knowledge and leadership of Aboriginal women has historically been obscured in government processes and Australian political debate. This is perhaps most poignantly demonstrated in the Hindmarsh Island bridge controversy of the 1990s, which centered on a site in the Coorong and Lower Lakes region in South Australia. In a protracted legal battle, Ngarrindjeri Aboriginal people fought to stop developers from building a bridge between the port town of Goolwa and Hindmarsh Island. Ngarrindjeri women Elders testified that the bridge should not go ahead for spiritual and cultural reasons. The Elders did not reveal details, as this would mean exposing sacred knowledge that was limited to some Ngarrindjeri women. The Elders' withholding of key details was widely debated in Australia, intersecting with broader politics about Aboriginal land and native title rights. This controversy generated the phrase "secret women's business," and conservative politicians and media framed the lack of information provided to the courts as indicating that the women were making false claims. Ultimately, the Hindmarsh Island bridge was built in 2001 with funds from the developers. However, the feminist anthropologist and consultant on the case, Diane Bell, has shown that the circumstances were far more complex. She argued that in this instance, and historically, protections around sacred women's knowledge clashed with men's dominance of the relevant bureaucracies, which in turn reinforced a systemic dismissal of Aboriginal women's knowledge in government processes. Anthropologist Deborah Bird Rose has called this kind of institutional process—in which the bureaucracies that are intended to decolonize actually perpetuate colonial imperatives—"deep colonising."[24] While steps have been taken by Australian and state governments in recent years to recognize women's knowledge and address institutional procedures that perpetuate colonizing processes, much more remains to be done. Within this context and history, this chapter aims to be attentive to the stories of women.

WEAVING IS GENERATIONS AND COUNTRY

As we drove along a rough dirt track, Meryl Mansfield-Cameron pointed out various plants amid the low scrub, calling out their names in Tanganekald with an explanation of what they were good for and also giving their names in Western scientific taxonomy. Meryl showed us *kengk muldi* / the smoking bush / bush coastal Olearia; another bush, *kundui* / sweet appleberry, with delicious berries; a rare parrot likes that shrub. The track we were driving along was part of the national park that encompasses the Coorong wetland. This long, thin, and saline lagoon is one of Australia's "icon sites" for the national government's Living Murray initiative and listed by the Ramsar Convention for Wetlands of International Importance.[25] We drove by a densely growing clump of sedges, and Meryl said, "That's the one I use for weaving.... See, it has the spines on the top and looks sort of like an emu's foot." The sedge was a mass of long, straight, bare stems, each of which had a tuft of growth at the top from which three sharp spikes protruded. These spikes were the toes of the emu's foot. I had asked Meryl about this plant before, called *bilbili* and *kayi* in Tanganekald, also known as *Cyperus gymnocaulos*, spiny-headed sedge or spiny sedge. Tanganekald is an

Aboriginal language group of this region, part of a wider area of Kukabrank Country, which encompasses the Coorong and Lower Lakes area. In the wake of British colonization in the 1830s, many Aboriginal people and groups in the area reidentified under the broader group name Ngarrindjeri, which has gathered significant support and traction.

Meryl is a skilled weaver and harvests the sedges at various points in the vicinity of the Coorong, drying them before using them to weave. These sedges grow at the edges of water and thrive in river and wetland environments across Australia except in the most arid parts and some tropical regions.[26] There is a long history of Aboriginal people harvesting the sedges for *lakan*/weaving along the Coorong, but many of the plants were only planted recently—some by Meryl and her husband Peter Mansfield-Cameron—as part of a revegetation program organized by the Aboriginal Lands Trust. The area had been cleared to create grazing paddocks for cattle and sheep starting in the late nineteenth century, and grazing still dominates parts of the wider area. It had also been denuded by rabbits, introduced to Australia in the late eighteenth century and then spreading across the Country.[27]

Meryl and Peter live on the banks of the Coorong on a property called Trunkeena, which they told me means "a duck sitting at the edge of the water" in Tanganekald. Peter's great-grandfather Alfred Cameron acquired the deed for the property as part of a settlement of debts. A European colonist had borrowed money from Cameron and, not being able to repay the debt, transferred the deed of the house to his name. This block constituted the ancestral lands of Alfred Cameron's grandmother, the most senior woman in the Temperumindjeri tribe. Thus both Alfred Cameron and Peter Mansfield-Cameron have lived on the lands of their ancestors. The history of Peter's family was told to them by his aunts, and their stories, collection of photographs, and family papers, as well as wooden and woven items that predate British colonization of the area, fill a room dedicated to the history of these generations, this place, and the Coorong. Meryl, a woman of Aboriginal descent from Millicent in South Australia, has over many years made Trunkeena and the Coorong her home. She learned to weave from Lettie Nicholls (née Karpany), a skilled Ngarrindjeri weaver. Meryl's weaving is prominent in the house and includes *koi*/baskets of many different shapes, from the taller slim-line sister baskets to wide, shallow baskets, as well as eel and duck traps made from sedges. All of these closely follow designs from the nineteenth century in this region but are likely much older

FIG 1.2 Small sedge basket woven by Meryl Mansfield-Cameron. Photo: Emily O'Gorman, 2020.

forms. Also prominent are egg scoops made from woven sedge, branches, paint, and gum nuts that are interpretations of traditional designs, and table coasters made from pine needles and raffia. Weavings, like lives and generations, are both dynamic and enduring.[28]

Design anthropologist Elizabeth Dori Tunstall has engaged with the practice of weaving as an important enactment of "being rooted." Aboriginal weavers often use the same kind of fibers and the same or similar techniques and patterns as their ancestors did across many generations, extending well into the generations that preceded British colonization of Australia. These techniques and patterns are passed on through specific family members or teachers and are adapted to incorporate new knowledge, networks, stories, and materials. Tunstall explained that "being rooted is different from being connected or even grounded. Grounding is only at surface layers. Being rooted goes as deep in the earth as above in the sky, providing greater stability."[29] Being rooted is not about being static and unchanging; it is about being strong and resilient, standing through generations.

Being rooted through weaving is shaped by the plants themselves, which flourish in particular Country. On the Macquarie Marshes, Danielle explained that "we've got two or three different types of reed in the marshes—and they have another reed that I've seen up in the Northern Territory which we don't have. You know, so you use your local material and that determines the pliability or how you might use it and the purpose you might use it for because of the character of that grass or reed." In the marshes, the sedges and reeds have particular characteristics that influence how they can be woven, shaping weaving practice and knowledge. Danielle explained that there are particular "properties of weaving with the sedge from the marsh, because you harvest it right, it's straight out of the ground, it's green, you weave it, and then as it dries it becomes tight on itself, and it has its own personality, I suppose. You know, you've got this item here, and it's green and lovely, and it becomes. Its natural form starts to influence how it is at the end."[30] The woven item *becomes* as the sedge dries and tightens. The weaver and the plant have together determined the new form but with a range of other influences—human and more-than-human—that have shaped the plant and the person.[31]

The importance of the properties of particular plants in shaping woven items means that in many places Aboriginal people have not only encouraged specific plants to grow in parts of wetlands and other watery places but also *to grow in particular ways* through burning, influencing the characteristics of the plants. Kate explained that burning sedges and rushes at the right time and in the right conditions, with water protecting the roots and a "cool fire," encourages the plant to grow new, stronger fibers, which create a different, stronger weave compared to fibers from plants that have not been burned in this way.[32] Women have been traditional custodians of Aboriginal fire knowledge at the fine-grained scale needed to encourage rushes and sedges for weaving. They have also been important custodians of men's landscape-scale fire knowledge, conserving this knowledge during the cultural upheavals of British colonization, including the active suppression of Aboriginal people's burning practices by colonial authorities. Because of the knowledge they hold, women have been central in revivals of burning practices in eastern Australia.[33] Kate and other Wiradjuri have started to reintroduce burning in specific land claim areas (a state government designation designed to recognize Aboriginal people's historic dispossession), working with Aboriginal people from Cape York in northern Queensland to test burning regimes on small patches of land. One of these places has a small dam where sedges and rushes grow, which they recently burned to change the properties of the plants. Kate said, "You don't want to burn it too hot.

You want to burn after it's flowered or seeded.... It had basically all died off, and there was water in there. So it wasn't a hot burn." Kate noted that this was a learning process for her and others involved—an experiment of sorts—so they are monitoring how the plants respond to burning and being left, in order to compare the burned and unburned plants and areas: "We did part of it, but we didn't do all of it, so we'll be able to see how it comes back." Kate is an installation artist and also weaves with these plants, "so I'm curious to see how different it is" to weave with the plants that have regrown after the burn.[34] Here it is important to note that Aboriginal fire practice and knowledge remain relevant in environmental management precisely because of this kind of adaptability and dynamism, which draws on knowledge formed across multiple generations and multiple, changing landscapes.[35]

On the Macquarie Marshes, Danielle explained that knowledge exchange both between generations and among different Australian Aboriginal cultures is an important aspect of her duties in caring for Country. Knowledge exchange strengthens roots. In the wake of often violent acts of colonization in many places in the Murray-Darling Basin in the nineteenth century, which continued into the twentieth century through government policies of assimilation, knowledge exchanges are important in caring for Country. Danielle said that "a lot of what I learn about my Country I've learned from other people's Countries, and they say, 'Now I've given you that knowledge, now you go home and apply it to your place.'"[36] Kate told me that in some places in NSW, like the town of Forbes on the Lachlan River, the state government's Local Land Services have facilitated the reintroduction of sedges along riverbanks that had been cleared. She said that along with planting the sedges, "they taught the women as well to do the weaving, so it was kind of mixing those two things."[37]

Sharing knowledge among women, including through and about weaving practice, is particularly important, Danielle explained, because there is cultural knowledge that is specific to women, and this is a way to "enhance and maintain and pass on cultural knowledge." It also helps to "build leadership confidence" among women.[38] For all of these reasons, Danielle has facilitated women's workshops on the Macquarie Marshes, including a large conference titled "Natural and Cultural Resource Management," which focused on weaving. This conference, held in 2012, brought together Aboriginal and non-Aboriginal women from across the Murray-Darling Basin. Kate and other Wiradjuri people from the Mudgee area participated in this workshop, exchanging knowledge about weaving and water. Kate explained that weaving was both the subject of the conference and the means through which knowledge was exchanged: "It was talking about the environmental issues around weaving, but it wasn't like with the white board and the PowerPoint, it was . . . weaving while everyone was talking for three days." The conference was a landmark event that facilitated important teaching and learning, and it was funded through Aboriginal organizations, Catchment Management Authorities, and other agencies sponsoring local Aboriginal women from communities. It had an explicit environmental focus, but as Kate told me, this cannot be separated out from cultural aspects, and in fact weaving ties multiple facets together: "You had people . . . who were there from an environmental perspective, and you had people there who were there as artists, so that perspective, and then there's sort of that cultural perspective, so they were all at the table, coming from different angles, but it was kind of like the weaving and the sedges . . . [brought together] all of the issues."[39] Similarly, Danielle emphasized to me that weaving cannot be separated out from other aspects of caring for Country, and there are other knowledges and practices that must be nurtured alongside weaving. She is currently organizing a workshop through the Western Aboriginal Women's Network, which will facilitate discussions among women knowledgeable in medicine, language, and weaving. Weaving is part of a broader set of connections and concerns, linking a range of knowledges, various practices, and spirituality. For many Aboriginal women I spoke to, weaving is a way of being on and caring for Country. Along these lines, Danielle said, "Weaving is one way of women being able to connect and utilize . . . resources . . . but this is a chance for us to give back . . . as well. We are looking at it from a different perspective."[40]

In particular instances it has been important for Danielle to share knowledge with men: "Men have men's business and women have women's, but we have to share aspects of our business to enhance culture, knowledge, and practice. I'll talk to men about something, and there's one meaning for me and another meaning for them, but when we come together, there's a third meaning, which is where it's a shared meaning. . . . It's a very collaborative approach to managing Country." In most Aboriginal cultures in Australia, men and women have different weaving knowledges and specific items to weave, some of which are connected to particular ceremonies or associated with specific duties. At times, if it is appropriate to share, points of connection can extend the meanings of each. In discussing weaving in the Macquarie Marshes, Danielle said, "Men would weave [items] like fish traps and nets.... If you look at some of the fish traps they made, they definitely look like pelicans' beaks. They'll do weaving with sedge, but then they might also do a weaving with sticks to have a funnel where the fish would be corralled into the funnel... like those eel nets that you see down in Victoria near Heywood. I've been out there with them,... and the men were a part of it. They made their particular implements, and we made our baskets for a different purpose but sometimes from the same materials."[41]

Yet sedge and rushes are not always available, for reasons of water as well as access. Meryl told me that in the Coorong area, while sedge is strong and best for many things, if it is unavailable, then raffia, strips of material, or something else can be used. For Meryl it is important to continue the practice of weaving with whatever material is available, as the practice of pulling the fibers toward you builds personal strength and attunement, *miwi*.[42] Danielle explained that in the Macquarie Marshes "weaving with sedge is highly prized because there's not so much of it. You can [also] do weaving with any other material like flax." She noted that weaving was important to continue even without sedge because of the way it facilitates important discussions. Weaving, she said, is "a practice through which we can pass on cultural knowledge, have that opportunity to be able to sit and gather, and talk and share.... If we only waited until when the sedge is ready, it's going to be less and less opportunities."[43] The opportunities will be fewer because of the way water continues to be managed by governments for irrigation interests in the basin as well as the effects of climate change. The latter is already intensifying dry and wet periods throughout the region.

Anthropogenic climate change means that droughts which lasted years have become longer, with even less rainfall, compounding the effects of dominant modes of water management in the basin that divert water flows from many wetlands to irrigation areas.[44] Kate told me about plans by weavers near Broken Hill, NSW—located toward the end of the Darling River system, near its junction with the Murray River—to place woven animals in the dry bed of the river to make visible the animals that should be there but are absent because of the lack of water flow, which has been diverted upstream, including to cotton farms near the Macquarie Marshes.[45] Weaving reveals interconnections between rivers and wetlands in the basin. Some Aboriginal weavers have recently incorporated materials other than sedges or rushes into their designs to draw attention to particular concerns, using weaving to show changes and problems in Country. For example, weavers from Arnhem Land in the Northern Territory recently incorporated fluorescent materials into traditional designs to show their concern over the environmental ramifications of mining in the area. Weavers in the Torres Straight, north of Queensland, have incorporated ghost nets—large fishing nets discarded or abandoned at sea—into their weaving to draw attention to the problems these cause for marine animals.[46]

Weavers interconnect the world, revealing changing politics and connections. From the late nineteenth century to the mid-twentieth century, making feather flowers was a common practice in missions in southeastern Australia. Christian missionaries aimed to assimilate Aboriginal people into European culture, often forcibly, and these institutions, along with governments, have been deeply implicated in the removal of Aboriginal children from their families. The practice of making flowers appears to have been introduced to Ngarrindjeri weavers by missionaries from Victoria, who then spread the practice to the Northern Territory.[47] In the 1920s, missionaries also appear to have introduced the coiled basket technique used by Ngarrindjeri people in to the Northern Territory.[48] While making feather flowers was common for a time, the kind of feathers that women used differed between regions. Ngarrindjeri women most commonly made feather flowers from black swan and pelican feathers, as these birds were numerous in places like the Coorong and adjacent lakes. Bell has noted that "the art of making feather flowers may have been taught by missionaries from Victoria, but men and women had numerous ceremonial and pragmatic uses for feathers and were already adept at working them."[49] They would sell these to passengers onboard steamships that visited Raukkan, then the Point McLeay Mission, located on Lake Alexandrina close to the Coorong. The Aborigines' Friends Association established Point McLeay Mission in 1859. The mission did not have enough resources or land to support the Aboriginal families who were relocated there, so selling feather flowers and woven baskets, along with fishing and farming, became important sources of supplementary income for many.[50] Design anthropologist Louise Hamby and Ngadjan Aboriginal arts curator Doreen Mellor have argued that "time spent on new forms, however, meant less time for the construction of classic forms that were used for ceremony as well as trade."[51] Nevertheless, Ngarrindjeri women continued to collect sedges and weave traditional forms. This was undertaken without the knowledge of the missionaries or the government, ironically aided by their and early anthropologists' dismissal of women's practices and spirituality. From the 1960s, new state laws that tightly restricted the hunting of pelicans made it more difficult to make feather flowers. However, the practice was passed along by women in some families, together with more traditional weaving, and it has been continued by Ngarrindjeri weavers such as Lettie Nicholls and Ellen Trevorrow.[52]

Historically, researchers have paid little attention to the practice of Aboriginal women's weaving in the Darling and Murray river regions. This is perhaps because it is often overlooked in historical documentary archives and anthropological studies, although there are passing mentions of women's mats and items for sale and some images from colonial observers. For example, in 1833 the surveyor and explorer Charles Sturt noted Ngarrindjeri women sitting on circular mats. These mats were drawn in 1844 and 1847 by colonial artist George French Angas.[53] Weaving has persisted, often through carefully guarded oral teachings that at times exploited British colonists' lack of interest in women and their practices, to avoid colonial censorship of this and other cultural practices.[54] At the same time, colonial collectors gathered examples of these items as artefacts, with extensive collections housed across Australia, including in national and state museums. These collections can be seen as different sorts of colonial archives that fail to account for the multiple meanings of weaving. Yet these collections can also be reinterpreted as showing the long histories of weaving in many places. Further, many families have woven items passed on through multiple generations, which can be an important source of information in contemporary practices. For instance, from her examination of a historical example of a sister basket made by the great-grandmother of her husband, Tom Trevorrow, Ellen Trevorrow was able to weave this design, interweaving knowledge across generations.[55]

Trevorrow, who was born at Raukkan in 1955, was part of a weaving workshop led by Ngarrindjeri elder Aunty Dorrie Katinyeri in the early 1980s. Trevorrow has spoken about the importance of this workshop in beginning her weaving practice, something she had seen as a child but not been taught—instead being encouraged to have a "white education." The decades since have been a time of cultural revival in Ngarrindjeri culture and Aboriginal cultures in many parts of the Murray-Darling Basin. Trevorrow has described the importance of weaving in passing on knowledge between generations, as collecting and drying the sedges and then sitting together to weave allow conversations to unfold and information to be passed on when the time is right. Trevorrow has used weaving workshops to show the importance of Ngarrindjeri culture and the Coorong to non-Aboriginal people. There, sedges are much less widespread than in her grandmother's generation, due to the intensification of upstream irrigation, the construction of barrages in the 1930s to support this irrigation by preventing freshwater from leaving the Murray River and flowing to the Coorong in dry years, and increasing salinization of the Coorong and nearby areas.[56] Far more than merely productive of artifacts or objects in museums, weaving reveals the importance of wetlands as socioecological places.

WEAVING IS WATER POLITICS

Weaving is tied in with water politics in the basin. Danielle explained the important connections between water, sedge, and weaving: "Weaving is [just] one practice, but it's reliant on water.... There's no water, there's no sedge." Many sedges and rushes grow at the edges of water bodies, and others grow nearby. Without these water bodies they die. Water and plants are Country, embedded within relations, including with people, and their absence has cascading consequences. Aboriginal women then cannot care for the plants, harvest them, or weave with these materials. With so many demands being made on the marshes and their water flows, Danielle has questioned whether she should harvest sedge during dry periods: "When there is sedge there [for weaving], I even wonder in these conditions whether it's even fair to go out and harvest because she's so depleted.... We have to look at resilience and how are we helping to sustain resilience in the landscape, in the marsh—and when there's been so much taking and clearing and everything, and then we compound that by going out and harvesting."[57] Water management in the Macquarie River system continues to compromise Wailwan culture and Country.

Aboriginal women have worked with water management bureaucracies to find ways to get water to wetlands. Here the sedges and rushes used for weaving have provided an important focus for negotiations with state and federal governments, as Aboriginal people argue for the cultural and environmental significance of the sedges and rushes so that water is redirected to areas in which they grow. Danielle has led this effort for the Macquarie Marshes, aiming to coordinate the release and arrival of water flows in the wetlands with the burning of rushes, reed beds, and sedges, so that the water fosters growth at the right times or protects the plants' root systems during a burn. The goals here are much broader, however. Danielle explained, "I suppose with government they want to see *tangible* things [like weaving and burning], and for us, our spirituality isn't always tangible. We're there because of our connection to the natural and spiritual world—then that enables us to practice our culture."[58] Weaving and burning in this sense are used by Danielle and others to translate Wailwan culture into something that can be more easily recognized and understood by dominant Western governments.[59] Yet weaving cannot be understood as merely the creation of artifacts nor as material culture. Danielle argues that weaving instead needs to be understood within a wider spirituality and set of practices of Wailwan people. Her statement draws attention to the importance of being on Country as fundamental to Wailwan culture. Weaving is being on Country.

While Wailwan people have sought out bureaucratic avenues to negotiate for water flows to rivers, floodplains, and wetlands, obstacles remain. Danielle explained, "When it comes to being able to say we want environmental or cultural flows for cultural practice, that's fantastic, but the reality is that the allocation might be there *or* [it might not].... Actually the old water sharing plans ... had 10 meg [i.e., megaliters] allocation for native title holders. They took that out on the sly. That was an allocation for the marshes."[60] Danielle currently represents Wailwan claimants in a native title application. In native title claims, which are often drawn out over many years, Aboriginal groups can regain some rights to Country by making claims in accordance with Australian federal law. In 2004, the Australian National Water Initiative, established to reform water governance, recognized the need for native title holders to have dedicated water allocations, which, as Danielle noted, are no longer available.[61] At the same time Aboriginal groups in the Murray-Darling Basin, particularly those represented within the Murray Lower Darling Rivers Indigenous Nations organization, have developed the idea of "cultural flows," which can be used to lobby governments for water allocations. This expands on the idea of environmental flows, included in government policy from the 1990s, which has also been used by Aboriginal groups.[62]

Environmental flows are sources of water managed specifically to meet environmental goals. These flows can be granted by governments in the form of water released from large dams if there is evidence that there will be significant environment benefits. Environmental flows are granted to particular organizations or individuals and can be one-off or recurring. Aboriginal people in the Murray-Darling Basin have negotiated for environmental flows or partnered with organizations like the Murray-Darling Basin Wetlands Working Group to combine these flows with practices of caring for Country that are simultaneously spiritual and cultural, making the case that these things cannot be separated. Here they argue that what is good for their cultures and spirituality is also good for the environment.[63] Danielle is involved in drafting a memorandum of understanding with a group of environmental water holders that formally recognizes these flows as simultaneously fulfilling cultural purposes, not only in the wetlands but also on the way to the wetlands, along rivers and in floodplains. The flow will thereby "enhance our ability to have cultural values [and outcomes] recognized."[64] Often conditions are attached to environmental flows, including that water will not be released if dam levels fall below a certain height. In dry times governments prioritize dam water to Western-style farms and crops, and environmental flows are often the first to be cut. There is in fact a history of environmental flows allocated for the marshes not being released from Burrendong Dam, or only small amounts being released, or water being illegally siphoned or diverted for irrigation.

All of these arrangements have many problems, including distinctions between environmental, cultural, and farming water that fail to recognize the interconnections between these categories central to Wailwan lore and law. In some instances the distinctions and hierarchies between these categories in Western culture, policy, and law can be used as extensions of colonizing processes that have sought to dominate lands, waters, and peoples. This is exemplified in the fact that Aboriginal people are not allowed any commercial gain from either environmental or cultural flows, and economic activities are barely mentioned in state legislation regarding Aboriginal water rights.[65] This is clearly a problem: the restriction singles out this group, limiting their financial capacities on the basis of *difference*. Danielle spoke about this injustice: "If we get a water allocation, we can't have any commercial benefits. . . . That's written in the law. We hope to challenge that in our native title case. I just can't see why we can't have it and everyone else does. We're not looking at it as a commercial thing where we'll go out there and clear the land—we actually look at it more like, How can we hold that water in a river or stream for prosperity of natural-cultural values?" The way this regulation currently functions unduly limits Aboriginal people's rights and financial opportunities. Danielle explained, "I think they're more worried about commercial gain at a bigger level. That we're selling our allocation, or cropping or grazing or something like that. . . . It's written in black-and-white there for us that we don't have any rights for that." All of these aspects shape wider water politics in the Murray-Darling Basin and limit opportunities for Aboriginal people to influence and be considered in water management. Danielle put this in powerful terms: "I just find that the current management in the Murray-Darling Basin is not inclusive," because lobby groups like cotton farmers exert a lot of influence while Aboriginal people have limited possibilities to do so.[66] At the state level, the Aboriginal Water Initiative was established by the NSW state government in 2012 in response to a growing recognition that Aboriginal people's values and knowledge needed to be better accounted for in water management, but it was abandoned in 2017. During its operation, this initiative, headed by a Kamilaroi man, Bradley Moggridge, made significant steps in developing mechanisms for engagement that recognized the cultural complexity of water for Aboriginal people and could provide an important basis for future similar work.[67]

Danielle moves between different spheres in representing Wailwan native title claimants, her employment by the NSW National Parks and Wildlife Service (NPWS), and her role on the Environmental Water Advisory Group for the Macquarie Marshes. While she balances these roles carefully, it means that she constantly needs to question how she presents herself as a Wailwan person. She said, "I'm really hamstrung. I've said, I am first and foremost a Wailwan woman.... They want to know that it's not impacting my work. You've got to be very clear ... when you speak out about the current water management regime and about what context you're speaking in, and you have to choose your words carefully." At the same time, working with government bureaucracy is essential for both getting water flows to Country at crucial times and gaining access to parts of Country that are under NPWS management. As formal avenues are removed or otherwise compromised and personal relationships become the only way to do these things, she finds herself at times in difficult situations, caught between different duties. Yet she sees it as important to be involved in government processes for a range of reasons, including because she is an Aboriginal woman.[68]

WEAVING IS ACCESS

Danielle has explained that "weaving is a way we open the door [to Country] ... to say that we as Aboriginal people want access for cultural practice."[69] Weaving here is a gateway to parts of Country that can be difficult to access for reasons of bureaucracy and land ownership, with some parts encompassed within grazing properties and other parts under NPWS management. Issues around access to the marshes have been ongoing since British colonization of the region, which was spearheaded by pastoralists in the 1830s. Pastoralists came from Wiradjuri Country to the south, following the rivers in search of grazing land. In Wiradjuri Country, frontier violence significantly escalated during drought and increased competition over resources in the period between 1838 and 1840, as pastoralists occupied Wiradjuri hunting grounds and sought to monopolize access to water.[70]

In colonial Australia, the importance of water to stock and people meant that rivers and wetlands were the first areas pastoralists sought out. As in Wiradjuri Country, frontier violence intensified in 1840 in Wailwan Country in the Macquarie Marshes region due to colonists' "unrelenting seizure of all rights and uses of land."[71] By the 1880s the marshes were hemmed in by pastoral properties, and in 1900 parts of the marshes, totaling forty thousand acres, were declared a game reserve by the NSW government. The government granted some pastoralists occupiers' licenses for these "marsh blocks," authorizing them to graze their stock there—predominantly cattle until the introduction of sheep in the area in 1927. By then many already acknowledged that grazing cattle in the wetlands was damaging the vegetation. These licenses became concentrated in the hands of particular farmers, a situation that remained until 1943, when the government moved to form smaller blocks for these licenses. Large sections of the marshes became a fauna reserve in 1955, with many of the birds listed as protected under state acts. Parts of the marshes were listed in the Ramsar Convention in 1986 and extended in 2000 and 2012.[72] Free-roaming cattle remain a concern for managers in the marshes, along with introduced feral pigs, foxes, carp, and weedy plants such as lippia that are well adapted to floodplains. All of these have flourished in the marshes, at the expense of other plants and animals, adding to the stresses of reduced water flows.[73]

From early British colonization to today, Aboriginal people's access to and use of the marshes have been limited in various ways, compounded by government policies of surveillance and forced removal, often to dedicated reserves and missions, a practice that extended from the colonial era well into the twentieth century, and arguably still continues. However, Wailwan people used a variety of strategies to continue connections to Country, including working at pastoral stations and living in camps along riverbanks and on traveling stock reserves, as well as Aboriginal reserves near rivers and wetlands. In many ways, contemporary debates about management of wetlands as socioecological places are part of ongoing disputes about Aboriginal people's legal rights to Country.[74]

Danielle told me that under state government laws Aboriginal people can access areas under the management of government bureaucracies, such as NPWS, for cultural uses, with the only restriction being on the use of threatened species. Nonetheless, in her experience access to NPWS areas has not always been easy: "It depends on the manager at that park at the time and their attitude as to whether you get access to sedge and [other] plants out there as well. Even though there's a policy and executive support—all of that—it's just the reality.... It's always at the discretion of the manager on the day whether you can get access or not."[75] In Australia, as elsewhere, in the twentieth century both private and government-protected areas have at times been managed along the lines of "fortress conservation," which in broad terms means creating and maintaining wildlife conservation areas by displacing or excluding human inhabitants.[76] This approach has also framed aspects of Aboriginal people's use of Country as at odds with species and environmental protection goals. Such approaches are rooted in colonial-era ideas, including racial hierarchies, and have been critiqued, challenged, and addressed in various aspects of protected areas and the wider environmental management process, including the reincorporation of Aboriginal burning practices in some private and government-managed conservation areas. Yet these historical legacies are strong enough that they continually need to be challenged.

At the same time, Danielle's employment by NPWS has at times facilitated her and other people's access to parts of Country, including the Macquarie Marshes. Further, being involved in government processes means that Danielle can directly challenge restrictions on access that are counter to government laws and policies: "I need to be there to be able to challenge that and say, no, that's not right. We have a policy, we have procedures around this. Aboriginal community can go out to Country to harvest resources." Harvesting sedge for weaving can provide a focus for negotiating access, but even so, Danielle has faced bureaucratic obstacles, needing to get a letter of permission from NPWS for the women's weaving conference in 2012. Danielle noted that all of these "little things"—the bureaucratic contradictions and minutiae of gaining access to Country—add up: "[They can] impact on your motivation sometimes. But it's not going to stop me.... You can get around that though very legal ways just through challenging that [initial decision]." Most importantly, for Danielle this means trying to make people think differently:

> Challenging the mind-set is the biggest thing, I think. It's also challenging the Aboriginal community's mind-set that just because they said no doesn't mean that that's the end of it. It's saying, "Well, we need to find an alternative way, but we can still access that." And that's what I feel I'm there for, to enable that and to speak for that and to provide opportunities of Aboriginal people to flourish on Country, and we determine what that is. We'll play within rules, but the rules also change all the time, and they need to consider that when it comes to our cultural practice as well."[77]

Relationships with landholders, especially pastoralists whose properties border the marshes, are also important in accessing particular areas, including for ceremonies. Danielle told me that recently she and other family members and friends scattered the ashes of one of her relatives on private land where her family knew the owner. Her deceased relative had wanted to "go home" to the marsh, so a place was selected where the water would come through when the marsh was healthy. Danielle also told me about a recent naming ceremony for a young relative that took place along the Macquarie River near earth mounds on private land. She said, "We're right in the heart of the marsh, still getting access through just a handshake agreement with the landholder there who we've had a good long relationship with," ensuring that "the next generation [is] connected with responsibility to maintain our connections to Country." In this particular instance, the relationship with the landholder stemmed from a large Aboriginal Network conference of about 140 people, which included weaving alongside other practices and knowledge exchange. This conference took place on NPWS land, but the landholders provided resources such as fuel when they were needed and were supportive of Wailwan people being on Country.[78]

Wailwan, like other Aboriginal people throughout the basin, continue to weave and maintain their responsibilities for taking care of Country, even when access is difficult or near impossible. Country, like weaving, is threaded into the world in multiple ways. Danielle explained, "Aboriginal women are still very connected to our roles and responsibilities from traditional law. We just have to adapt in the role because we can't get on Country. Our purpose is to nurture Country, people, and culture." Weaving and knowledge about Country continue to thread generations together, making worlds, including water and climate change futures: "Despite the conditions out there, I think we want to maintain our connection, but we're also reviving and renewing and sharing our culture with our younger ones too, to connect them for the future."[79]

TWO

LEAKING

Containment and Recalcitrance of Swamps

THE SIGN TOLD ME I WAS APPROACHING KEARNEYS SPRING HISTORIcal Park. Catching sight of a windmill, I made my way up a small rise. Closer to the windmill, water seemed to be seeping out of the ground in multiple places, forming small channels of clear, flowing water. It was difficult, however, to pinpoint the precise points where water was coming from. Along one channel were the remains of a well, which pooled water and appeared to be augmenting the flow. Nearby, an old horse trough, once fed by the windmill's pump, leaned precariously against support posts. This leaky place is a headwater of what is now known as West Creek, which runs through the city of Toowoomba, located on the Great Dividing Range in southeastern Queensland. Today, West Creek is a channelized watercourse, with parklands lining its banks. This creek and nearby East Creek and their parklands were once part of swamps extending along the lengths of these watercourses.

These swamps have been central to the city's history. Starting in the mid-nineteenth century, European colonists sought to control and contain them as a source of disease and damaging floods, while relying on their aquifers for water supply. This chapter considers the swamps as historical actors, paying particular attention to their "leakiness" and recalcitrance, as

FIG 2.1 Parts of Toowoomba's swamps have been drained by the council since the late nineteenth century in an effort to "improve" them by making them productive or recreational places. Pictured is a drained section of the eastern swamp that is today a parklike recreational area. Photo: Emily O'Gorman, 2017.

they troubled borders, boundaries, and townships—characteristics revealed within and against ongoing attempts to contain and control them. It focuses on the period from early British colonial settlement in the 1850s to the 1940s. During this period the Toowoomba Council undertook significant, ongoing drainage works, among other actions, that aimed to "improve" and civilize the swamps by controlling and containing them. The swamps instead leaked out with a kind of a spatial and temporal promiscuity. They moved underground though aquifers and through the air via miasma and mosquitoes, beneath and out into residential and commercial areas. Their leakiness, then, was a sign of good conditions for some. The swamps could also reappear as floodwater, inundating homes and businesses—something they still do today, as demonstrated by large floods that swept through Toowoomba in 2011. Used as grazing land, dumping grounds, and for market gardens, the swamps were fluid actors that defied the colonial council's attempts at controlling and civilizing both people and nature. Indeed, issues of class and race became important in these efforts to civilize the swamps, as the council sought to limit use by poor white colonists and Chinese market gardeners.

This chapter particularly focuses on disease as one of the key ways in which the swamps leaked. Across this period dominant Western scientific understandings of disease transmission changed significantly, at various times encompassing miasma, germ theory, and mosquito-borne diseases. These changes shaped the council's persistent but shifting attempts to control the swamps. Some of the hallmarks of the council's approach to the swamps—its focus on improvement and the valuation of watercourses as water supply, all the while regarding them as wasteland—resonate with histories elsewhere in Australia and around the world.[1] This was, after all, a period of intensive swamp drainage in many places. However, these universal logics played out in, and were challenged and reshaped by, particular places and people in different and specific ways, cocreating particular more-than-human worlds.

The term "leak" is more than an appropriately watery metaphor. Leaks draw attention to boundaries and attempts to both contain and keep out. They can be transgressive, unsettling, and monstrous.[2] Elizabeth Grosz has examined leaks in relation to bodily fluids, which complicate understandings of solid physical boundaries and of cleanliness, while also drawing attention to corporeal interconnections.[3] In examining the leakiness of Toowoomba's swamps, we can start to appreciate the interconnections of the swamps with the world beyond and how these far-reaching relationships unsettled colonial desires for the swamp to be a defined and controllable space and also troubled a range of other boundaries, including human bodily boundaries through disease. This in turn draws attention to the roles of multiple organisms and forces in coconstituting the swamps—from rotting organic matter, to bacteria, to a range of animals, to the rocks containing aquifers. This chapter, then, continues to reveal these and other watery places as lively human and more-than-human places.

In researching the agency of the swamp as well as colonists' attempts to contain them, I have drawn on multiple archives and other historical materials, such as newspapers, because the swamps have seeped into many narratives, perhaps a trace of their characteristic leakiness. In these sources, colonial ideologies and power relationships are evident. They also reveal some of the many human relationships that have shaped these histories, including the roles of Aboriginal people, Chinese immigrants, and European colonists, as well as nonhumans such as mosquitoes.

A TOWN BUILT ON SWAMPS

It is unclear how long Toowoomba's swamps have existed. They run through and permeate volcanic basalt soils that formed between 27 and 19 million years ago. A series of violent volcanic eruptions likely formed this part of a mountain range, now known as the Great Dividing Range, creating multilayered lava fields that covered over sandstone. The swamps formed at the top of an escarpment of this range and have supported and been shaped by many animals, plants, and people with dynamic histories. Aboriginal people have lived in this region, interacting with and shaping the swamps, for thousands of years. In 2002, a heritage report tried to reconstruct the swampland terrain of the early nineteenth century, conjuring a landscape that was quite varied, in places containing open eucalyptus forests and, along parts of the watercourses, the floodplain river red gum, with Cumbungi reeds growing in the deeper water holes. The swamps, fed by underground springs and rainfall, formed a combination of watercourses, lagoons, and marshy areas, supporting an abundance of freshwater fish, yabbies, waterbirds, and other animals. Aboriginal people's burning regimes shaped the swampy landscape, parts of which were likely burned regularly. This is the Country of the Giabal people, although many other Aboriginal groups have camped and hunted in the swamps, especially during large gatherings at the nearby Bunya Mountains. For these gatherings, which happened every two to four years until the late nineteenth century, people would travel from up to 250 kilometers away for ceremony, trade, and meetings, which centered on bunya nut harvests. The swamps, located on the pathway to the Bunya Mountains, provided an opportunity to rest, exchange information, hunt animals such as ducks, and catch fish. The swamps, connecting underground and surface, aquifers and rainfall, are also part of the song lines of Giabal people and those of other Aboriginal groups in Australia, which extend in multiple directions, linking waters, mountains, and ancestral spirits.[4]

The history of Toowoomba as a township is steeped in European colonial pioneer narratives of pastoral expansion, a period of violence and upheaval for Aboriginal people across eastern Australia but often narrated by European colonists and their descendants in heroic terms.[5] The pastoralists came from the south in the 1820s and 1830s, bringing with them sheep and cattle to establish illegal pastoral stations that the colonial government then legitimized through surveys. These pastoralists initiated a period of intense frontier violence in the region. Towns soon followed to service the needs of the stations. In this region, which colonists called the Darling Downs, conflicts between Aboriginal groups and pastoralists continued from the 1840s to the 1890s, often centering on access to land and water, as well as conflicting views about animal ownership and use. In this context, in the 1850s, a small colonial settlement was established at the swamps in Giabal Country.[6]

In the late 1840s, colonists used the swamps, located just eight kilometers from the colonial settlement of Drayton, for timber supply, thatching, and as a "camping ground for passing drays."[7] A shepherd from a nearby pastoral station had set up camp there and was making charcoal near the western swamp.[8] Traveling British colonists trying to find a way to cross the swamps witnessed Aboriginal people hunting waterbirds in thick reeds and applying mud to their skin; the colonists assumed this was to keep the many mosquitoes from biting.[9] At this time the swamps were, in many ways, outside the reach of expanding colonial powers and regulations, used by Aboriginal people, ex-convicts, and colonial travelers, as well as a range of animals, plants, and insects.

In the 1840s, frontier violence between Aboriginal people and colonists in the wider region intensified, and in many cases Aboriginal people retreated from areas that were becoming centers of European settlement. Aboriginal people may therefore have sought refuge at the swamps. Here and elsewhere in Australia, some Aboriginal people worked on pastoral stations, as stockmen and domestic workers, and others lived in and around the townships that were being established. Though colonists often portrayed them as "outsiders" living "on the fringe," more recent histories have sought to recenter "fringe camps," examining the roles and strategies used by Aboriginal people to stay on Country in the midst of widespread dispossession and the nuanced interactions between local Aboriginal people and colonists both on farms and in towns.[10]

As Drayton's population expanded and water supply ran low, colonists began to look for an alternative site for their businesses. In 1850, the NSW colonial government surveyor James Charles Burnett laid out town allotments at "the Swamp," as it became known to colonists (the region remained part of NSW until Queensland's separation as a colony in 1859). The strict grid pattern placed the center of the town in the heart of the swamplands, nestled in the crook of East and West Swamps, as they were later called. Where the swamps met, they created one large swamp, known to colonists as the "Main Swamp." This large swamp then drained into a watercourse called Gowrie Creek, named by colonists after a nearby pastoral station.

FIG 2.2 Toowoomba was built on and alongside swamps. Colonists called these East Swamp, West Swamp, and Main Swamp). Detail of map from 1857 showing Toowoomba swamps. Plan of Toowoomba at Drayton Swamp, Parish of Drayton, County of Aubigny shewing the lands recently measured for sale, circa 1/1/1857–circa 31/12/1857, Department of Natural Resources and Mines, Qld 2017, Queensland State Archives Item ID714269, Map.

Increasingly, colonists moved their businesses from Drayton toward the better-watered swamps.[11] By 1855, the population of Toowoomba was eighty-four, and Drayton sixty-one.[12]

Although the Swamp—also called Drayton Swamp or Toowoomba Swamp—possessed resources that colonists needed, it also had negative associations. For example, colonial traveler and entrepreneur Nehemiah Bartley, recounting his journey through the area in July 1854, called it "green and oozy" and quickly continued on to "old Drayton," also known as "the Springs."[13] "Springs" and "swamps" held very different meanings for the mostly British colonists, who brought with them understandings from their homelands. In Britain in the mid-nineteenth century, springs were associated with health and cleanliness. Swamps, by contrast, were frequently seen as sources of disease and danger more generally, long associated with menacing spirits and beings.[14] These negative associations may explain why colonists did not seek out the swamps earlier. Colonists may also have largely avoided them because they continued to be used by Giabal people and may have been a stronghold of theirs in the frontier wars. Elsewhere in Australia, similar kinds of watery areas avoided by colonists became refuges for those who were discriminated against in colonial regimes, including Aboriginal people and escaped convicts.[15] The name of the Swamp was officially changed to Toowoomba in 1857–58 to shed the negative connotations of swamps. The etymology of Toowoomba is contested but most likely derived from a Giabal word for the swamps or a Giabal interpretation of the English word "swamp".[16]

As more settlers bought land and moved to the swamps, the area eclipsed Drayton as a business and population center, achieving incorporation as a municipality in 1860, with Drayton eventually becoming one of its suburbs. By that year, the population of the Darling Downs included 7,000 colonists, 1.5 million sheep, and 140,000 cattle.[17] An advantage that Toowoomba had over Drayton was its closer proximity to the new Toll Bar Road, completed in 1855. This road became a key route linking the Darling Downs with colonial settlements east of the mountain range (primarily Ipswich and Brisbane). The Toowoomba swamps became a main en route watering place for horses and cattle as drays and herds of livestock were driven through town.[18]

The swamps shaped other aspects of early settlement. Flooding of the swamps' watercourses became an almost yearly occurrence, inundating nearby residences and businesses.[19] Floods also changed the quality of the swamps' water supply, making it "very hard and aluminous," an issue that was somewhat overcome by the sinking of public and private wells for additional water supply.[20] Although European colonists generally considered higher land more desirable and healthier, underpinned by a belief that low swampy areas led to ill health and racial degeneration, the importance of surface and well-water supply meant that in Toowoomba the blocks neighboring the swamps became the wealthier areas, and poorer residents were forced to purchase land farther out and on ridges.[21] However, at least in some places, poorer and often marginalized groups lived along the swamps.

MIASMA, "IMPROVEMENT," AND STINKY SWAMPS

With greater numbers of people, new industries, and concentrations of new kinds of animals, the swamps changed character too. Colonists felled many trees for building material. Cattle grazed and drank at various points along the swamps, which had become crown land and were treated as a kind of common by residents, stockmen, and dray drivers.[22] Animal hooves would have created muddy bogs and, along with colonists' land clearing, facilitated erosion. Soon the swamps started to expand. Simultaneously, the shallow channels of the swamps were used by those in the town to dispose of waste, including animal carcasses, rotting vegetation—seemingly including swamp plants such as reeds along with dumped vegetable waste—and slop water, which, in the words of one colonist, combined to form "noxious exhalations," understood by many at the time to be miasmic vapors that caused disease.[23] The rotting flesh of animal carcasses and decomposing plant material combined in putrefaction with other organic matter to create odorous vapors, which colonists thought might be deadly.

Residents traced ongoing outbreaks of scarlet fever and diphtheria to the swamps, and in the mid-1860s, the Toowoomba Council began to lobby the new Queensland government for permission and funding to drain them for sanitary reasons and reclaim some of the swampy ground for productive use. In arguing the case in 1865 in the Legislative Assembly, William Henry Groom, the member for Toowoomba and Drayton, noted that along with rotting waste in the channels of the swamps, "the main swamp in Toowoomba was a sort of natural cesspool for the district. In the summer the reeds in it grew something like seventeen feet in length, and these drying down and rotting at the end of the year, caused a great amount of effluvia to arise and diffuse itself over the town." He argued that it was because of "residence in swampy localities" that the local infant mortality rates from typhoid, diphtheria, dysentery, and other diseases exceeded even those of Brisbane, notable for its high infant death rates.[24]

Miasma was part of a medical theory prominent in the nineteenth century that disease was spread, as noted by historian Stephen Halliday, by "inhaling air that had been infected by corrupting matter," often associated with odor and places like swamps.[25] As was evident in some of Groom's statements, miasma was widely understood as having a seasonal effect. In swamps, bogs, and a range of water bodies, these odors were understood to be emitted by rotting vegetation and a range of other sources, causing deadly diseases such as malaria, dysentery, and typhoid. The use of these localities as dumping grounds in places like Toowoomba seems to have both been precipitated by and reinforced this association. That is, people's view of swamps as miasmic wastelands seems to have meant they were inclined to use them to dispose of unwanted matter, which exacerbated their miasmic qualities.[26]

Some colonists who owned or rented property created gardens along the edges of the swamps and planted trees such as willows in an effort to transform them from "wastelands" into places of "production" and to be more aesthetically pleasing. Groom noted that "very large sums of money had been spent improving the land in the vicinity of the swamps—in one case he knew that as much as £400 had been spent on clearing and planting with vines and fruit trees a single garden but on account of the swamp in the vicinity increasing, the garden had to be given up and now the bottom fence was completely hidden from view, the swamp having encroached upon the ground so far." The colonists' efforts were undermined by the swamps, which continued to grow in size, encroaching into the adjacent blocks. In one case, the swamps had expanded to "within eighteen feet of a man's house."[27]

In many ways, improvement was the dominant environmental ideal of the nineteenth-century West, a pre-preservationist model that assumed all nature needed human action to perfect it. Historian W. K. Hancock has pointed out that the notion of improvement, widespread in colonial Australia, assumed that the places of "the new country" were somehow lacking and needed to be made more like the "the old country," particularly England.[28] Improvement was frequently enshrined in law; to maintain possession of land colonists need to build fences, cut down trees, and drain swamps. Therefore, as historian Tom Griffiths has argued, nostalgia for Britain was not innocent: "Improvement . . . was dismissive of indigenous environmental systems; aggressive as well as possessive."[29] In its legal and cultural forms, improvement was an essential part of colonizing processes that simultaneously dispossessed Indigenous people of lands and waters and sought to reconfigure environments into something more familiar. It also held a moral imperative whereby both environments and people could simultaneously be civilized through hard work and healthfulness.[30]

For the Toowoomba Council, it was a drawn-out process to gain legal control of the swamps and funding for drainage works, which had been stalled by various legal and financial complications. With the swamps continuing to change in shape and character, Groom thought that the longer the engineering works were delayed, the more expensive they would be. In 1867 he began to lobby the Surveyor General of Queensland. In making his case, Groom enrolled the expertise of "medical gentlemen" who "affirm that the Swamps in the present condition, receiving daily as they do, large quantities of decomposed vegetable matter, remains of dead animals and other deleterious substances are calculated to form hotbeds of disease. The disease of dyphtheria [sic], so prevalent among children and so fatal in our town, as many as thirty two in the first quarter of 1866 having died from it, arise[s] chiefly from the noxious exhalations of these unhealthy localities." Typhoid and other disease outbreaks were also frequent.[31]

People's activities brought new diseases and increased the area of the swamps. Yet there were clearly other organisms and elements that had agency here too, including the vegetation of the swamps and animals. Their behavior and biology helped shape the swamps. There is little detail given about the animals that grazed the swamps, although horses and cattle were among them. Their eating habits and size, the fleshy decomposition of animal remains, along with plant growth and decay, shaped the character of the swamps.[32]

To support their case to the surveyor general, the council commissioned a report by engineer Hugh Swann. Swann set out extensive drainage works, which included the construction of some small reservoirs in the swamps for mixed use, including providing domestic water supply and watering cattle. One of these he recommended be in a depression already "completely hollowed out by the brickmakers" in the East Swamp (south of today's Queens Park) because the water there was "good and clear." Swann claimed that with this scheme, the swamps would be "made pure and wholesome" and become "very valuable" instead of "despised."[33]

One of the great promises of such hydraulic engineering for many at the time was transforming environments like these into healthy, productive, civilized areas. In making his case for extensive drainage works, Swann drew on examples of other swamps that had been transformed through drainage work, like "the Fens in England" and, more locally, the "scrublands" of Oxley Creek and the Pine River near Brisbane. Here and elsewhere, engineering schemes of this sort were part of colonial enterprises that simultaneously reflected, created, and reinforced class divisions. As a warning about not draining the swamps, Swann gave the example of Frog's Hollow in inner Brisbane, which was "notoriously the most unhealthful spot." This area flooded easily and had become the settlement's red-light district. It was notorious for bad drainage and loose morals; here "unhealthiness" took on an incivility and uncouthness. There was a sinister snobbery and moral surveillance at work in understandings of swamps. In examples like Frog's Hollow, the poor had been left to live in the damp miasmas while wealthier areas were drained. In Toowoomba the class dimension was more apparent. Indeed, it was perhaps because a number of wealthy people had bought land adjacent to the Toowoomba swamps—and Groom highlighted this in various forums when he explained the garden "improvements"— together with its ongoing encroachment into the town, that the council continued to advocate for its drainage. Groom argued that the cost to the government of the drainage works would negligible, as the drained land could then be subdivided and sold for £20 per acre, which would defray the costs.[34]

The surveyor general of Queensland, Augustus Charles Gregory, in principle supported the drainage of the swamps for sanitary reasons. This was, however, "provided that the ground be either cultivated as gardens or used as grazing paddocks, though the swamp's drainage would be of doubtful advantage if the land were left as a waste common where all description of rubbish and nuisances might be deposited." Animals and plants, then, were central to the draining of the swamps, not just as a cause of their enlargement and odor but as a way they could be made productive as grazing paddocks or gardens. Gregory considered there to be two main issues with the council's proposal: first, that the land would probably not sell for the amount the council proposed "on account of its unhealthiness... and ... [as it would be] subject to occasional inundation," so the costs of the drainage would not be recouped; and second, that landholders who had initially purchased blocks adjacent to the swamps would no longer have access to this water supply. The matter rested.[35]

Despite concerns about these drainage schemes, the Toowoomba Council persistently lobbied various Queensland government departments, noting that the swamps continued to grow in size, cause disease outbreaks, flood, and be "useless" instead of "productive." It was not until 1873 that the government began to acquiesce. That year, Gregory—still the surveyor general of Queensland—surveyed the swamps to provide a description to the

Public Lands Department and determine the extent of the drainage works. For these bureaucratic bodies and engineering works, the precise nature and extent of swamps needed to be clarified. However, before Gregory got far in his account of the swamps, he encountered a problem: "The term swamp has been applied in such an indefinite manner that it is difficult to determine what is required." Local colonists and earlier surveyors had simply called these areas swamps, but to what and where, precisely, were they referring? Where did the swamps start and stop? Rather than the word alone being imprecise, we might see the swamps themselves as somewhat imprecise and having participated in their own ambiguity. The swamps, in their swampy, leaky way, seemed to defy the specific descriptions sought by the colonial surveyor; they instead seeped out into the town. They also defied and transgressed the straight lines of the original survey, at times impinging on, and under, the blocks; they had changed significantly even in the short time since the town was established. Ultimately, Gregory wrote that the term "swamp" "has been assumed to indicate the whole of the vacant space between the surveyed allotments and the roads within the boundaries of the municipality." The swamps, which had been the reason for the settlement, were now merely the "vacant space" in the map between the surveyed allotments.[36]

In 1874, following Gregory's survey, the council gained legal control of the swamps, borrowing £2,000 from the Queensland government to undertake the drainage work. Work began the same year under the direction of an engineer, James Thorneloe-Smith, and was completed by 1875. The project drained stretches of the eastern, western, and main swamps that bordered the center of the town, the engineer noting that they were made up of "spongy turf" formed by "annual peat deposit." The works aimed to both dry out parts of the swamps and mitigate floods. Thorneloe-Smith recommended that after the second phase was complete, "the lands on either side of the main drain be enclosed by a fence of some sufficient description to keep cattle from damaging the works already completed"; in this way he addressed the council's concerns over the effects of cattle in expanding the size of the swamps.[37]

The council began leasing parts of the reclaimed swamps soon after the drainage works were completed, for such uses as residences, businesses, and the grazing of animals. This was against conditions placed by the government on the council's use of the swamps. The government had temporarily granted the council control of the swamps so that it could drain them to improve the health of the town, but this did not extend to financial benefit.

In fact, at any time the government could take back control of the swamps, as they remained crown land. Yet the council continued to assume it had complete ownership of the swamps and in fact in 1883 sold some of the drained swampland back to the government via the Railway Department. The attorney general described this as "so much waste of paper" during an inquiry several decades later into the council's financial gains from illegally leasing and selling the swamplands.[38]

Despite Thorneloe-Smith's confidence that the works were adequate to mitigate floods, a sudden large flood in 1876 showed that they were not. While one observer claimed the flood to be "the highest ever seen in Toowoomba," putting a town bridge "four feet under water," uprooting trees, inundating buildings in the town center, and washing away carts and other possessions, subsequent floods also defied the drainage capacities of the new works and leaked into the town.[39]

Leaks are points of connection as well as breaches in containment. They reveal relationships that are undesirable or unwanted in some way, to at least some people. Floods revealed interconnections with other water bodies, including networks of creeks and rivers, showing that the swamps not only leaked into the town but, farther out, they likewise received leaks. The council had tried to control and contain the connections of the swamps to these water bodies. But this was not easy to achieve, as the watercourses found the swamps again, spreading into the town and seeping into aquifers, at the same time replenishing well-water supplies.

CESSPITS, WELLS, AND UNDERGROUND SWAMPS

Surface water drainage aimed to address the visual and olfactory issues of rotting animal carcasses and vegetable matter. However, as frequent typhoid, dysentery, and other disease outbreaks continued, the townspeople began to turn their attention underground, to the aquifers that lay beneath the surface of the swamps. In the late 1870s, the close proximity of wells to cesspits stirred suspicion among some residents, including medical practitioners, that contaminated sewage was leaking into aquifers tapped by wells, which were a significant source of drinking water in the town. Toward the end of the nineteenth century, more people began to draw connections between sewage-contaminated drinking water and poor health. Importantly, these connections were made during the cholera epidemics in London in 1854 and 1866 and then in Hamburg in 1892.[40] In many ways, the public health responses in Toowoomba to typhoid, dysentery, and other enteric diseases from the 1870s through the 1890s chart a shift in the dominant approaches to disease in Western science, from miasmatic to germ theory. Yet they also show that miasmatic theory was still influential in this period and often subscribed to, in tandem with germ theory, by medical professionals and others.[41] Further, this new interest in water supply as a source of disease led Toowoomba's residents to reconsider the hidden depths of the swamps, particularly the interaction of wells and cesspits through underground water tables.

Prior to Thorneloe-Smith's construction of the drainage works, typhoid, dysentery, diphtheria, and other diseases had recurrently broken out in the town and were together blamed for its high infant mortality rate. The council and townspeople hoped the drainage works would reduce the miasmas that caused disease. Yet all of these diseases continued to occur in the town, and following another typhoid outbreak in 1878, the council assembled a commission to investigate reasons for ongoing outbreaks of disease. In May of that year, the commissioners reported their findings.[42] The members of the committee were William Groom; John Thomson M.B., a medical practitioner; and a prominent builder in the town named John Garget. The recommendations and responses of the council and commissioners, invariably an elite group of white colonists, served to reinforce stratifications of class and race. Those in power especially targeted poorer white colonists and the animals they kept, along with Chinese people and their farming practices, as sources of disease. At the same time they sought to control and civilize the swamps.

The report began by noting the impact of typhoid and other "enteric disease" (into which category the authors also placed diphtheria) on the mortality rate in Toowoomba: by the end of April 1878, there had been forty-one deaths attributed to enteric disease, accounting for "36% of the total of all deaths" in the town that year. From there, the commissioners concluded that "the relative situations of the two (wells and pits) make it imperative that there should be a connection between them, and it cannot be otherwise." Many blocks of land in the town backed on the swamps, which were dotted with gardens and stables, as well as cesspits and wells. The commissioners conservatively estimated that there was a ratio of five hundred cesspits to three hundred wells in the municipality, although, they noted, it was likely that each of the seven hundred to eight hundred dwellings had a cesspit that collected waste and needed to be regularly emptied. According to the commissioners, at least some citizens showed "carelessness" and "inattention" in maintaining their cesspits, which were left to fill up and then overflowed during heavy rains. This account was supported by the nightman hired to empty cesspits, who gave evidence that only fifty cesspits were emptied each year, "and many of them had not been disturbed for five or six years." The commissioners also thought a recent drought had caused the contents of the cesspits to become concentrated, as they were not diluted by rain. The stagnant sewage then leached into the aquifers, contributing to the disease outbreaks. The nightman buried the contents in the East Swamp, near a campground (the current location of Queens Park), but soon after the commissioners' report was published, a resident alleged that he had not done this properly, thereby possibly contaminating the well-water supplies nearby. The commissioners asserted that most of this had been attested to by "medical witnesses." In the sciences, reliable and modest witnessing was widely seen as crucial in arguments for these correlations, a gendered role occupied mostly by men viewed by other scientists as rational and objective. The commissioners also gave weight to "the opinion of the few townspeople who very wisely have given up the use of well-water," including medical doctors. Even as the council tried to drain the surface water, the aquifers tied people to the swamps. They were the underground swamps, which continued to leak into dwellings and into bodies via the wells.[43]

Further, the commissioners argued that the disease outbreaks had also been caused by "total neglect of ordinary cleanliness" and "ignorance." They made a point of describing "overflowing cesspits, filthy pig-styes, dirty poultry houses, offensive middens, sometimes dry but more frequently wet and consisting of every possible kind of refuse, putrefying accumulations of fruit and vegetables, ill-kept drains, and stagnant slop water and slime." All of this, the commissioners asserted, fostered "disease and death," which "might most naturally be expected" in the town. The commissioners made ten recommendations, including replacing cesspits with dry earth closets (similar to contemporary composting toilets, but the contents were not used as compost) and using reservoir or rainwater instead of well water. They also recommended "the banishment of all pigs within the municipality" and "the supervision of Chinamen's gardens."[44] The comments and recommendations of the commissioners—three white males in professional positions of power in the town—resonate with arguments by a growing middle class in other cities in Australia, Britain, and many other places in this period for slum clearance, based on the view that the lower classes had poor sanitary conditions.[45]

The removal of pigs from the municipality aimed to address two concerns: the possible transmission of typhoid and hog cholera from pigs to people and the offensive material associated with the feeding of pigs. Here we can see the mixing of germ and miasmatic theories, as disease transference between pigs and people relied on germ transmission while the odors from rotting pig feed as a cause of disease were in line with miasmatic theory. The commissioners wrote that the pig feed often consisted of "animal or vegetable refuse and scraps," which if uneaten by the pigs were "left to decay just where they have been cast," creating strong smells, long associated with disease.[46] Piggeries were water intensive and located near the swamps, contributing to an understanding that these areas were generally unhealthy.[47] The poor often reared pigs, and thus this recommendation pathologized the practices of a particular group of people as well as animals. Excluding pigs from the municipality would significantly impact this group, who would need to change their practices or move to the outskirts of town. Indeed, similar health controversies over pigs in urban centers such as New York also manifested along class lines.[48]

Aside from poor pig farmers, the other socially stigmatized group that was singled out was Chinese market gardeners. As in many towns and cities in Australia at this time, the market gardens farmed by Chinese migrants in Toowoomba were an essential source of fresh fruits and vegetables. By the 1870s, there were many Chinese market gardens along the East and West Swamps, particularly toward their junction with the main swamp.[49] The ambivalent feeling of European colonists toward the swamps may have meant that Chinese people could work parts of these areas without others seeing them as competing for valuable land. In fact, in many parts of Australia, market gardens farmed by Chinese migrants were often located on swamplands or riverbanks close to towns for reasons of water supply, but perhaps also because these places were not sought out by white colonists and were often only semiregulated by councils and governments. At times stereotyped as only growing cabbages, Chinese people working the gardens grew a range of vegetables and fruits, likely introducing species and varieties from China such as bok choy and Chinese radishes, which would have become new features of the swamps' vegetation. In Toowoomba, Chinese farmers sold their harvest to Chinese and non-Chinese customers from carts, hawked produce door-to-door, and sold it in Chinese-owned shops in the town. The Chinese market gardeners of Toowoomba cleared, drained, and irrigated parts of the swamps, adapting technologies developed over thousands of years in southern China.[50] Historian Joanna Boileau has argued that Chinese migrants shared with European colonists a "desire to 'develop' the land and make it productive," and white settlers begrudgingly admired them (and perhaps sometimes became jealous of them) for doing so with such success.[51] The swamps provided a source of income for Chinese growers, but they were also dangerous, as large floods, such as in 1893, could sweep away produce and belongings.[52] Chinese market gardeners shaped the swamps, while the swamps shaped their lives and livelihoods.

The horticultural practices of Chinese market gardeners included the use of various kinds of animal manure, including that of pigs, as fertilizer.[53] The recommendations of the commissioners regarding pigs and the monitoring of Chinese market gardens may well have been linked, although the commissioners' allegation that farmers used human excrement as fertilizer in the gardens was a key focus of this section of their report. The commissioners argued that an inspector should oversee the storage of excrement, both animal and human, and "the manufacture of fluid manure," which they regarded as "a considerable danger to health." In a statement resonant of germ theory, they claimed the liquid fertilizer was "rained over the plants themselves" and that "solid particles" were likely ingested by customers when eating lettuce, salads, and "cooked vegetables." The commissioners recommended that the council "prohibit the further establishment of these gardens in such close proximity to thickly populated portions of the town."[54] At this time, anti-Chinese feeling was increasing among those of European descent in many parts of Queensland, and more broadly in Australia and the Pacific World, and the sanitary conditions of the market gardens became subsumed within broader ideologies of race.[55] The exclusion of Chinese people's market gardens from the town potentially had important consequences, including social isolation. It also meant more labor would be required for the Chinese gardeners to bring harvests into town, where customers and shops were concentrated.

These recommendations for disease prevention aimed to refigure human and nonhuman relations, partly by excluding particular people and animals from the town. They were part of a hygiene-focused colonialism, racism, and classism through which colonial authorities sought to establish social power and environmental control. Historian Warwick Anderson relatedly argued that "excremental colonialism offered a potent means of organizing a new teeming, threatening environment."[56] Ultimately, the commissioners urged that "the sooner all excretions, whether fluid or solid, whether of men or animals, are removed from around our dwellings, the better, and yet instead of removing them or destroying them we hoard them up as if they were treasure, and they putrefy and ferment, and a concentrated decoction percolates through the soil to contaminate our wells, while a poisonous gas is given off to pollute the atmosphere and so what between the water we drink and the air we breathe, we are charged with the germs of disease and death."[57] This statement further draws attention to the commissioners' simultaneous adherence to aspects of miasmatic and germ theory.

These ongoing disease outbreaks were at odds with Toowoomba's growing reputation as a healthful place to visit. Located in a mountain range, with cooler weather than the more tropical coastal areas of Queensland, Toowoomba was reputed to have a climate more amenable to European constitutions. This view was steeped in the ideology that particular climates suited particular races, and those of cooler climates were more civilized. Toowoomba was becoming a retreat for the wealthy white colonists from Brisbane and Ipswich.[58] The council's many actions to prevent disease, including swamp drainage and the commissioners' recommendations to exclude particular animals and people from the town, can be seen as attempts to maintain Toowoomba's reputation as a healthful and ideal place for wealthy white visitors. As the commissioners wrote in the conclusion of their report, "It must be noted that Toowoomba has long enjoyed the reputation of being a healthy, salubrious town. For years past it has been free from contagious disease of every kind, except in 1875, the measles epidemic, and the recent outbreak of typhoid."[59] This, we know, was not entirely true, as there had been previous outbreaks of typhoid, dysentery, and other diseases. Such misrepresentations might be explained as a kind of propaganda directed at visitors, since this report was published in full in the local newspaper. Souvenir images attempted to portray a clean, civilized town that featured health, prominent buildings, and gardens. There are in fact very few images of the swamps (with the only exception being images of floods in the main streets).

In response to the typhoid epidemic of 1878, the council moved to establish a piped water network (as opposed to well water), which was completed in 1880.[60] The council introduced and gradually extended a dry earth closet system.[61] However, the condition of the swamps, including the state of the drains constructed in 1874, was still cause for complaint by residents.[62] The swamps continued to make their presence felt in uncomfortable ways for the townsfolk, particularly as a source of large floods in 1893 and 1906 that damaged infrastructure such as roads, bridges, and in the case of the latter flood, also businesses and houses.[63]

The council's efforts to civilize the swamps, Chinese gardeners, and poor people were echoed in the Queensland Government's efforts to civilize Aboriginal people. By the turn of the century, the Queensland Government had assumed control over Aboriginal people in Toowoomba and throughout the Darling Downs. Government Protection Boards frequently removed people who remained on their traditional lands to reserves and missions located outside of town and often away from their traditional lands. In this period, removal from Country made performing traditional ceremonies impossible; these practices were also suppressed through legislation.[64] In the words of Jarowair Elder Paddy Jerome, whose Country includes the Bunya Mountains, "The bunya festivals stopped when they removed us."[65] This was a deliberate attempt by the government to sever Aboriginal people's connections with Country and civilize them.

Chinese market gardeners remained in Toowoomba well into the twentieth century but continued to be blamed for disease outbreaks. This was much longer than other towns and cities in Queensland, where growing anti-Chinese sentiment among white colonists in the late nineteenth century pushed out many Chinese market gardeners.[66] Animals and their relationships with the swamps also continued to be a source of concern for the spread of disease.[67] However, in the early twentieth century, the focus of public health shifted to mosquitoes.

MOSQUITOES

In the late nineteenth century, scientific proof of the life cycle of the malaria parasites and the role of mosquitoes as vectors in transmitting that disease to humans resulted in a surge in research into possible diseases spread to humans by these insects. In the twentieth century, mosquito elimination became a key aspect of public health campaigns. The link between disease and mosquitoes changed people's relationships with mosquitoes and the watery places in which the insects bred. In Toowoomba, which was declared a city in 1904, it led to renewed efforts to drain and otherwise control the swamps.[68] Such approaches were evident in many places, but in Australia,

Toowoomba's specific strategies became a model adopted by other towns and cities.

The success of the Toowoomba Council in destroying mosquitoes was largely attributed to a local physician, Dr. Thomas Price, who came to be known as "the Mosquito King." From 1912, Price advocated for a mosquito survey of the city by the state government.[69] Toowoomba, like other places in Queensland, experienced outbreaks of bubonic plague traced to rats and fleas and of mosquito-borne diseases such as dengue fever.[70] These vectors thrived in close proximity with humans, and public health campaigns aimed to eliminate their favored habitats, which were often places cocreated with, but neglected by, humans. Like Toowoomba's swamps, these were often seen as dirty places that troubled the firm boundaries many people tried to establish between humans (and human-controlled places) and nature.

In 1914, mosquitoes were particularly plentiful in Toowoomba. Early that year, local businesses joined Price's call, urging that "steps . . . be taken to combat the present mosquito pest," and in 1915, the Queensland Government agreed to a mosquito survey of Toowoomba. This was undertaken by L. E. Cooling from the Department of Public Health. Cooling found that the most common mosquito was *Culex fatigans* (now called *Culex quinquefasciatus*), a carrier of filarial, a parasitic worm that he thought likely bred in the swamps and town water storage. The next most abundant was *Stegomyia fasciata* (now more commonly called *Aedes aegypti*), which had been introduced to Australia in the 1880s as it spread around the world from Africa through shipping networks.[71] This mosquito could carry yellow and dengue fever, and Cooling thought that in Toowoomba it mostly bred in outdoor water cisterns and other small pools of water near human dwellings. Females of both species of mosquitoes fed on humans and during feeding could transmit diseases. Cooling surveyed the mosquitoes in September and noted that although "the mosquito nuisance at present is slight . . . a different state of affairs must exist in the summertime" (December to February), when the mosquitoes bred in greater numbers, as higher temperatures were needed for larvae to hatch.[72]

The swamps featured in Cooling's survey of *Culex* mosquitoes and his descriptions reveal not only that these mosquitoes inhabited these places but also the broader changes to the swamps that had fostered mosquito breeding sites. When reporting on *Culex* breeding places, he wrote, "Toowoomba is traversed by what are locally called swamps. . . . A 'drawn-out cesspool' is perhaps the most applicable name for it receives the whole town sewage and storm water in what may be termed from a sanitary engineer's point of view 'the combined system.'" In dry weather "crude sewage" filled the watercourses, "in which larvae of *culex fatigans* delight to breed." He thought that "these sewage streamlets are the principal mosquito breeding place of Toowoomba." The East and West Swamps, after receiving local waste, fed into Gowrie Creek, which Cooling described as "really a slowly moving body of town sewage." After leaving the town, this creek gathered waste from several factories, where butter, milk, and bacon were produced. At the points where this waste entered the creek, Cooling described a putrid stench and ideal *Culex* mosquito breeding grounds, where many of the larvae were present. The factory waste provided a nutrient-rich growing environment for the larvae. Cooling described the swamps in the town as having "luxuriant... vegetation, including rank swamp and other grasses, weeds as well as green and other coloured algae and fungi," and he used the different algae and fungi to determine the levels of pollution and filtration. The "streamlets" of the East and West Swamps were silted with mud and debris where cattle and horses wandered, creating mosquito breeding places with their hoofprints.[73] People's relationships with the swamps, as grazing grounds and places of waste disposal, had helped shape ideal breeding places for mosquitoes, "as if for them."[74] We might see this as a feedback loop of sorts for modernist schemes, which are reinforced as their consequences play out.

Cooling's *Stegomyia* survey revealed another aspect of the town. Although the water supply was "pipe borne," people still supplemented this with well water, which was "lifted to elevated storage-cisterns through the agency of windmills." In these cisterns or water tanks, which were uncovered or partially covered, he found *Stegomyia fasciata* larvae. They were also in rainwater tanks and likely in the street cisterns for public use as well as water-logged gutters. These mosquitoes had little to do with the swamps, at least directly. Rather, the smaller pools of water in and around human dwellings and buildings were ideal larvae breeding grounds (some of which came from the swamps' aquifers), with a supply of human blood nearby for female adults.

Cooling recommended a range of "temporary measures" to quickly limit general mosquito breeding in the town. His first recommendation was "rough canalisation" to deepen the "streamlets" and clear them of vegetation "so as to give the sewage a clear run." This was not only for mosquito prevention, he noted, but also for "general sanitary benefit." This primarily seems aimed at *Culex* mosquitoes. Second, he recommended oiling "all stagnant and moderately running water" with a crude petroleum oil or kerosene-based spray once a week, a proposal that may have aimed to counter both *Culex* and *Stegomyia* breeding. Ultimately, however, Cooling argued that what was required for a reduction of *Culex fatigans* was "a sewage scheme" in the town, which would need to be constructed for more thorough and permanent results. A sewerage system that conveyed waste in pipes was started in the center of Toowoomba in the 1920s and gradually extended into other parts of the city.[75] In addition, Cooling recommended that the reservoirs be stocked with fish that ate mosquito larvae, such as native crimson-spotted sunfish (*Melanotaenia duboulayi*) or carp gudgeon (*Hypseleotris* spp.), which were already present at the headwaters of the West Swamp. To stop *Stegomyia*, he recommended that all tanks be screened, cistern openings covered, and wells boarded over. Further, he recommended that all tins at a local rubbish tip be buried and the ground leveled to stop the pooling of water.[76]

As a result of Cooling's survey and recommendations, the council established a Toowoomba Rat and Mosquito Board in 1916–17 under the Queensland Rat and Mosquito Prevention and Destruction Act of 1916. This act aimed to facilitate control and eradication measures for these two animals, both then seen as significant threats to public health. At the same time, Toowoomba's council boundaries were redrawn and the City of Greater Toowoomba created. At these elections, Dr. Price campaigned for a seat on a platform emphasizing health, especially mosquito control to combat dengue fever, arguing that the cost of treating this and other insect-borne diseases was more than the cost of preventing them. So central was mosquito eradication to Price's campaign that his rivals derided him as "the Mosquito King."[77] However, his message of vector control, disseminated through advertisements in theaters and film reels in public meetings, convinced many. He was successful in the election, appointed chairman of the Health Committee, and then elected mayor in 1918. Price established systematic programs of mosquito eradication for Toowoomba. The council regularly oiled the swamps and water bodies or sprayed them with distillate and removed vegetation. Screening tanks became compulsory, with fines for breaches and various other regulations. These were enforced in house-to-house inspections. The council also established an education campaign, the wartime expense of which was justified by Price as important in maintaining healthy citizens.[78]

Following the end of the Great War, Australian state governments monitored returning soldiers for malaria, and when one case was identified in Toowoomba in 1920, the Queensland Government urged the council to put into effect mosquito eradication regulations "without delay" in case of an outbreak. The city inspector—often a person with public health experience—was already making regular reports on the mosquito and other work. In one report made in 1920, he notified the council that he had discovered the larvae of *Anopheles* mosquitoes, which could transmit malaria from an infected human to other humans. He wrote that "seeing that there are some malaria patients in the town it is essential that we keep this species under control." A case of dengue was also reported in 1921. With so many mosquitoes identified as vectors in Toowoomba, and more and more species linked to disease transmission by scientists, mosquito elimination campaigns targeted all mosquitoes as a precaution.[79]

The city inspector's reports reveal a ramping up of efforts to contain the swamps starting in the 1920s. This included patrolling the border between the swamps and its animals. Hoofed animals were banned from walking in parks and areas like the swamps so that their hoof marks did not create mosquito-breeding places. This also aimed to mitigate the spread of ticks, and the council impounded unattended animals and fined owners in breach. The city inspector oversaw a series of incremental drainage works that aimed to further remove moisture from the swamps. All these regulations and drainage works again reconfigured various human and nonhuman relationships.[80]

Animals participated in this refiguring in sometimes unexpected ways. For instance, fluctuations in mosquito behavior and numbers, influenced by various factors, including rainfall and temperature, made eradication programs difficult to implement in a consistent way. These dynamics in mosquito behavior meant that inspectors had to adjust their efforts and behavior accordingly. It also meant that mosquito numbers could rise quite suddenly. For example, in 1938–39, mosquito numbers increased significantly, described by one resident as "hoards." Residents urged the council to increase its efforts to suppress the "pest," reporting on the location of "infestations."[81]

In 1920, in a paper delivered at the Australasian Medical Association conference held in Brisbane, Price advocated for mosquito eradication and prevention to be undertaken more systematically throughout Queensland, as many towns were "badly infested." As in his Toowoomba campaign, Price justified the need for action in terms of the relative costs: prevention was cheaper to the state than treating the diseases, which could become endemic, as with malaria in parts of north and northwest Queensland and filaria in Brisbane, and create epidemics, as with a dengue outbreak in 1904. He further noted that "yellow fever was an ever threatening menace."[82] He ended with a series of recommendations—essentially that the process he developed in Toowoomba be adopted all over the state.

In 1923, the state's hookworm campaign employed entomologist Ronald Hamlyn-Harris to undertake a mosquito survey of the Darling Downs, focusing on the mosquito-transmitted parasites of filaria and malaria. This was part of a broader set of public health activities undertaken by this organization to target parasites. In his report, Hamlyn-Harris singled out the practices in Toowoomba as being the best in the region at preventing mosquito breeding. He wrote, "Situated as it is on what was once a swamp of considerable size, Toowoomba lends itself particularly to mosquito breeding—and though Toowoomba is now practically free from this pest it is only by constraint and vigilance that the City can keep this enviable position." Despite his enthusiasm for the processes of eradication of mosquitoes in Toowoomba, Hamlyn-Harris found *Culex quinquefasciatus* near the center of the town and *Stegomyia aegypti* (*Aedes aegypti*) "in defective tanks a mile or more out of town." He also warned that the hydrology of the East Swamp, fed by springs and "scoured out" by storm water to form pools, created conditions for *Culex quinquefasciatus* to breed if left unattended. The water holes would need to be continually filled in and sprayed with oil. Hamlyn-Harris also found *Anopheles* breeding in part of the East Swamp, and here troughs, depressions, and "low lying paddocks would soon become infested unless cleaned and oiled weekly." Again here the absence of fish predation was key. He praised the drainage work that had turned a part of the swamps "so wet and boggy that cattle could not graze on it" into a paddock through spade-cut drains, which were also oiled. Outside the city limits, mosquitoes were abundant, indicating both the city's success in reducing mosquito numbers and the danger to humans in these areas, albeit being less populated. He concluded his report on Toowoomba by stating, "The responsible members of the Toowoomba city council are to be congratulated upon their work and its results."[83]

Toowoomba soon gained a growing reputation as being free from mosquitoes and the diseases they carried, which supported its claim as a healthy place to visit. This reputation was fueled in part by newspaper reports of the city's success, which detailed the methods used by the council, a scheme perhaps aided by the wealthy rate base of a booming pastoral district. The methods included regularly removing any receptacles that gathered water, tarring over holes in the streets, screening tanks, clearing vegetation from watercourses and treating weekly with oil, and extensively draining swamps. Although Toowoomba had already undertaken significant swamp drainage, the mosquito campaign motivated further efforts, and as one newspaper reported, the council set out to concrete a section of the swamps' channels each year, filling in the neighboring swampland as it went. Gradually, the swamps became concrete channels. From the mid-1920s to the mid-1940s, these council projects drew on workers from employment programs during the Depression. A number of councils in Queensland and New South Wales, including inland places like Quirindi and coastal areas like Maitland, wrote to the Toowoomba Council for further advice and details on the methods it had used to eradicate this "pest." Toowoomba became an exemplar for mosquito eradication, and the methods used by the council were implemented elsewhere. The city inspector's reports distilled some of the principles that underpinned these efforts into slogans: "No water, no mosquitoes"; "No mosquitoes, no dengue."[84]

The new focus on insects, particularly mosquitoes, as vectors of disease renewed efforts to drain swamps in many towns and cities in Australia. In the 1930s, Queensland Government regulations for "mosquito prevention and destruction" required all swamps and other water bodies on public or private land that, in the opinion of the local authority, might breed mosquitoes to be either drained or "effectively treated" with oil to asphyxiate the larvae and prevent adult mosquitoes from laying eggs. The outbreak of World War II, far from directing attention away from mosquito eradication, refocused attention locally, nationally, and internationally.[85] Techniques and approaches changed significantly during and after World War II, including the spraying of DDT in homes and mosquito breeding places like swamps. The range of sanitizing activities undertaken by some residents and the council can be seen as attempts to enact a firm boundary between the town and the swamps—more specifically, human bodies (and at times domestic and agricultural domains) and swampy more-than-human worlds. They were attempts to control human citizens as well as the swamps. These attempts intersected with and enacted ideologies of race and processes of colonization, linked closely to understandings of civilization and imperatives to improve environments.

While visiting Toowoomba in 2017, I tried to walk as much of the swamps as possible. The watercourses are now mostly channelized gutters of concrete or stone, with the swampy banks transformed into parkland. In the city center, at times the swamps' channels formed the edge of a parking lot or disappeared from view into underground drains with highways above. The swamps have remained both backwaters and parts of an idealized European landscape. However, they continue to assert themselves, leaking out temporarily and spatially. In 2011, large floods inundated not just the parklands but many places in the city, replenishing aquifers and damaging buildings. Farther outside the city, I visited the Toowoomba Waterbird Habitat that forms part of the eastern swamp, now called East Creek. There I met with one of the people involved in its creation. The project had been funded as part of Australia's bicentennial celebrations in 1988, which celebrated two hundred years of British colonization. It was a poignant move on the part of the field naturalists to try to restore part of the swamps at a moment when the process of colonization that had radically reshaped them and dispossessed Aboriginal people of their land and waters was widely being celebrated. In 2001, the Toowoomba City Council commissioned a heritage report on the city's waterways. As part of preparing the report, the consultants contacted Aboriginal Elders, and with local historians and others, they walked around the swamps together. They talked about the songlines of the swamps and the stories contained in places. The Elders showed the group the headwaters of the swamps: the springs feeding East and West Creek, something that none of the others had known.[86] In turning toward, rather than away from, the swamps, the heritage consultants and council reevaluated the Aboriginal history of the area. Yet there is also a danger in turning toward the swamps and Aboriginal people at the same time, as this can conflate Aboriginal people with the natural environment, which has a long and problematic history.

As we walked around the Waterbird Habitat, my guide told me that the area had been a paddock at the time they received the funding. Unable to find much information on how the swamps had changed, the designers and managers just did their best. We walked farther, and my guide pointed out the flood debris pressed against the fencing. Australian white ibis had moved into the islands in the lagoons to nest but were unwanted by the surrounding human residents because they were noisy and smelly. We found the remains of a well that my guide thought had watered a Chinese market garden. The constructed ephemeral lagoon did not work out as intended, as it did not fill frequently enough. The casuarina trees were not doing what the managers wanted them to, taking up more space on the banks of the lagoon than planned.[87] The swamps still leak, spatially and temporally, unable to be contained or controlled, and always recalcitrant.

THREE

INFECTING

Irrigation, Mosquitoes, and Malaria in Wartime

PALM TREES LINED THE WIDE STREETS AS I DROVE TOWARD THE CENter of the town of Leeton, located in the Murrumbidgee Irrigation Area in southern inland New South Wales. The grass that covered the ground near the road was lush and green, in stark contrast to the dry bush I had just passed through. The town's old water tower was close to the main street, built on a slight rise for practical reasons but also serving a symbolic purpose as a monument to modernist water management. Built in the early twentieth century, the MIA was part of a wider move to develop irrigation by Australian governments at that time, so as to water the inland and make it more productive of particular foods and goods. The NSW government built this scheme to harness water from the Murrumbidgee River and wider Murray River system for intensive agriculture, and it is today supported by a number of large dams. The MIA was a manifestation of ideals of control over water and indeed nature. Yet these ideals have been undermined from the beginning by soil salinity, water logging, drought, and as this chapter highlights, mosquitoes.

The pulses of wet and dry years in eastern Australia shape possibilities for life. From the late nineteenth century, the significant redistribution of water in the Murray River system for irrigated agriculture changed these possibilities in particular ways, with mixed consequences, including new interspecies dynamics. In the late nineteenth century scientific proof of the life cycle of malaria parasites (protozoans from the *Plasmodium* genus)—with certain stages occurring within *Anopheles* mosquitoes and in hosts such as humans—also spurred greater research into this and other mosquito-borne diseases. It also changed people's relationships with mosquitoes and watery landscapes, including irrigation areas. Malarial fever had been commonly associated with the miasma of watery places, especially in tropical and subtropical climates. Indeed, "malaria," derived from an Italian word for bad air, had also been known as swamp fever.[1] With this new research, watery places were no longer just dangerous to go to. They could come out into the world, onto farms, and into homes via mosquitoes, who were the unwitting intermediate hosts and vectors of parasites that could maim or kill humans, transferring the disease as they moved between infected and uninfected people.

Today, malaria is not readily associated with the MIA. Yet between 1919 and 1945, various state and national government departments coordinated at least five separate investigations into the possibility of a malaria outbreak there. This concern was sparked by the settlement in the area of soldiers returning from the Great War and World War II and researchers' discovery of an *Anopheles* mosquito (*A. annulipes*) in the region that could act as a vector of the parasite. Many feared that the combination of infected humans and mosquito transmitters brought the MIA dangerously close to an outbreak, with newspaper headlines such as "Malarial Mosquitoes: All Over Irrigation Area" (1929); "Malaria Menace" (1929); and "Malarial Dangers on Murrumbidgee Irrigation Area" (1946).[2] The NSW Water Conservation and Irrigation Commission (WCIC), which oversaw the MIA in this period, created an archive related to these investigations.[3] This archive reveals some of the fascinating ways knowledge and imaginings of nonhumans are world-making, and how particular imagined ecologies have taken shape. It further reveals the roles that mosquitoes and malaria parasites played in cocreating these imagined ecologies.[4] Underpinned by powerful political and social goals, irrigation areas could not be drained or easily oiled, as many swamps and other watery places were, presenting possible new medical, environmental, and ideological challenges.

I use the term "imagined ecologies" in an effort to bring together understandings with their world-making possibilities and consequences. "Imagined" does not mean nonexistent, nor do imaginings somehow exist

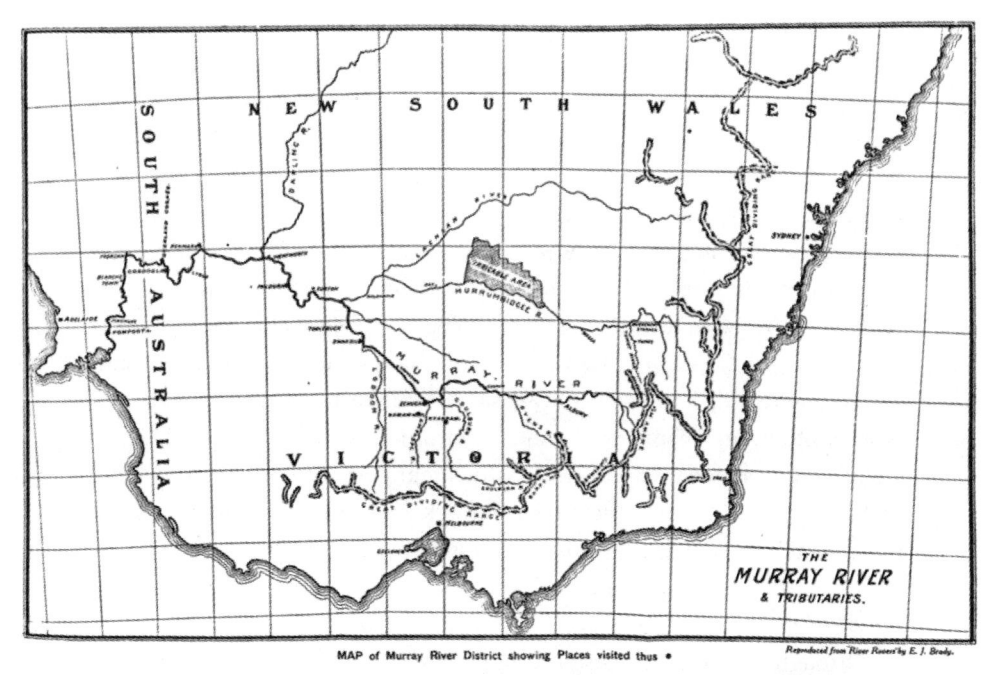

FIG 3.1 Map from F. H. Taylor's report into the presence of Anopheles mosquitoes in the Murray River system, showing the places he visited while undertaking the survey in 1916–17 and the location of the Murrumbidgee Irrigation Area (shaded section marked "Irrigable Area"). Credit: F. H. Taylor, *Malaria Mosquito Survey* (Melbourne: Commonwealth Government of Australia, 1917), "Malaria Mosquitoes—Murrumbidgee Irrigation Area," 1917–1946, Item 45/7990, Series 14511, Murrumbidgee Irrigation Area [MIA] files [Water Resources Commission], 1911–1986, State Records NSW.

separately from material environments. Rather, knowledge and understandings are always situated and partial, yet they guide actions and in so doing help shape particular kinds of worlds.[5] "Ecologies" brings attention to relationships that have been coconstituted, including those between humans and nonhumans. I do not intend to read back the concepts and approaches that now constitute the field of ecology but am rather referencing particular aspects of its meaning and history. Drawing on anthropologist Anna Tsing, approaching the past in a relational way allows us to see humans within temporally as well as spatially dynamic "webs of interspecies dependence"— in this case, webs in which humans, mosquitoes, malaria parasites, and many others were caught.[6]

Historians of medicine have traced the emergence of the field of disease ecology to the interwar years, with some tracing it even farther back, to early parasitology in the late nineteenth century, and specifically early malariology.[7] As Warwick Anderson has argued, such investigations drew on traditions of natural history that made them more "biologically animated" and complex than medical geography and more focused on disease evolution and mobility than laboratory investigations.[8] Further, historian Linda Nash, focusing on the Central Valley of California in the nineteenth and twentieth centuries, has shown that an examination of understandings of health and diseases like malaria across this period, and their "inescapable ecologies," can complicate histories of ecology that begin after World War II as well as enrich examinations of human attempts to control and exploit environments. This is because diseases have long intimately intertwined people, other organisms, and wider environments in ways that demand attention to relational connections, which many have attempted to control in various ways.[9] Indeed, these relationships were precisely what interested the researchers undertaking the investigations into malaria in the MIA, although they did not couch their investigations explicitly in terms of ecology.

Through research on mosquito-borne diseases, which intersected with and influenced public health agendas in Australia and elsewhere, people learned to see environments differently—that is, in terms of mosquito behavior and elimination.[10] Wetlands had long been drained, dried with plantings of trees such as eucalyptus, or altered in other ways for fear that they were a source of miasma.[11] That any pool of water could be a breeding place for deadly mosquitoes meant that eliminating water in the landscape was undertaken with renewed urgency. In many places, pools were also oiled, mostly with kerosene, and later sprayed with DDT. Parasites and, by proxy, mosquitoes—as well as other nonhuman targets of public health campaigns like rats, bacteria, and others—shaped these responses, which often aimed to exploit particular patterns in their behavior. Yet these nonhumans were often responding to changing human settlements that created favorable environments and closer proximity.[12] In coconstituting each other, these humans and nonhumans were acting out Anna Tsing's observation that "*human nature is an interspecies relationship*" that shifts historically as organisms, interactions, and understandings alter.[13] As many nonhumans became targets of sustained elimination campaigns, humans cocreated their lives in new ways too, with far-reaching consequences.

Mosquitoes became caught up in new imaginings of environments and human health (and their interrelationship) in the MIA by researchers, farmers, managers, and public health authorities, in the context of both changing landscapes and shifts in scientific knowledge of malaria in the twentieth century. Further, these imaginings were shaped by increasingly militarized approaches to environments, and mosquitoes, in the context of two world wars. These imaginings intersected with fluid, but ecologically and politically potent, categories that aimed to establish firm human control over agricultural areas and demarcate temperate from tropical, homes from nondomesticated nonhumans. Whether it is the boundaries we draw around ourselves as individuals or around communities, nations, and international organizations, who and what is seen as belonging are far from fixed. Instead they are created, with important consequences in these "contact zones."[14] These boundaries were held in tension with the intimate ecologies of the MIA, of people and mosquitoes living closely with and being constituted by each other in changing environments.

Initially centered on simply identifying the presence of *Anopheles* mosquitoes in the region, these investigations gradually became concerned with whether *A. annulipes* were in fact effective transmitters of malaria. These mosquitoes were a mysterious presence in the MIA in the early twentieth century, as they still remain. This history is intimately tied to irrigation, as governments tasked researchers with building up an understanding of the mosquitoes' distribution, abundance, and behavior in and around irrigation areas, which encompassed farms as well as towns. Their research into and knowledge of the mosquitoes were closely bound up with the landscapes of irrigation and imperatives of settlement. The power and politics of knowledge are also evident, as those who oversaw the running of the WCIC and MIA dismissed scientists' advice and local concerns over malaria, in many instances doing little to prevent an outbreak. This added to the imagined ecologies of the region, to an ongoing fear that a malarial outbreak would occur.

Little scientific research was undertaken in this period into other kinds of mosquitoes in the region, beyond some hasty observations of other species by scientists as they surveyed *Anopheles*, some of which were later recognized by scientists as vectors of other diseases (such as Barmah Forest virus and Murray Valley encephalitis). This may be due to the applied nature of governmental scientific research in this period, as resources were often concentrated on specific problems. It may also reflect the MIA's seeming lack of will to engage with the mosquito research, perhaps not wanting to bring further negative publicity to the area. The regulatory and governance structures of the state and nation, along with local politics and world events, all played roles in shaping knowledge of mosquitoes and malaria and, in turn, the imagined ecologies of the MIA.

MALARIA, MOSQUITOES, AND WAR

The World Health Organization declared Australia to be free of malaria in 1983, but prior to that the disease was widely considered to be endemic. Although it is uncertain whether malaria's endemism in Australia predated colonization, it certainly seems that colonial settlements and trading created favorable conditions for outbreaks. The history of malaria in Australia is commonly associated with the tropical north. In 1843, Port Essington, on the Cobourg Peninsula north of Darwin, was devastated by malaria; it was abandoned by 1849 because of continued outbreaks. In 1866, fifty of the seventy-six residents of Burketown near the Gulf of Carpentaria died from what was later thought to be malaria.[15] Yet there have also been instances of local transmission of malaria in temperate regions, including Sydney and Bega in NSW.[16]

In the early twentieth century, concern over malaria often led to public health campaigns that targeted mosquitoes. However, scientific evidence for the connection between the parasites and insects was relatively recent. In 1898 Ronald Ross, a British medical doctor working in India, demonstrated that mosquitoes transmitted avian malaria, and the following year a team of Italian malariologists showed that human malaria was transmitted by a local species of *Anopheles*. Later posted to Sierra Leone, Ross demonstrated that *Anopheles* also transmitted malaria to humans there. In general terms, for transmission to occur, a female *Anopheles* mosquito needed to bite an infected human, the parasites needed to develop in the mosquito, and then the mosquito needed to bite a receptive human host. In each place the *Anopheles* species was different, but a general understanding emerged that *Anopheles* mosquitoes were vectors of malaria. In the twentieth century *Anopheles* became the target of campaigns for the eradication of malaria around the world, being anthropomorphized in public health campaigns as criminals or femmes fatales, often in terms that referenced racial stereotypes. Particularly in the mid-twentieth century, these portrayals intersected with the racial stereotypes and language of war.[17]

The understanding of the roles of both people as carriers and mosquitoes as vectors that emerged from the late nineteenth century created new

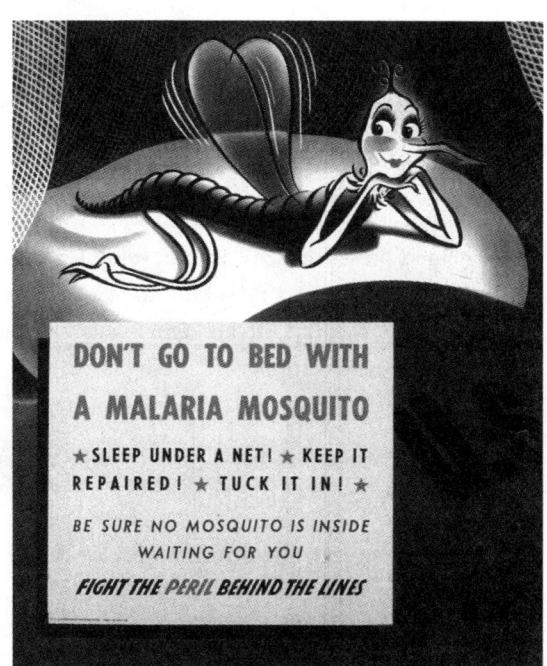

FIG 3.2 An antimalaria poster produced by the United States government during World War II, which depicts an Anopheles malaria mosquito as a femme fatale. Credit: "Don't Go to Bed with a Malaria Mosquito," US Government Printing Office, 1944, US National Library of Medicine.

perceptions of the possible mobility of malaria. Returned servicemen became a source of medical concern as possible carriers of the disease during and immediately following the Great War (1914–19).[18] During the Great War at least 1.5 million soldiers from all combatant armies were infected with malaria, forcing evacuations of servicemen and weakening the military capacity of many national armies.[19] The Australian Army faced its first epidemics of the disease in New Guinea in 1914–15 and in Syria in 1918. Medical personnel in the Australian Army became familiar with malaria during this war, and a number continued to contribute to malariology after they returned to Australia.[20] With an understanding that infected ex-servicemen could only cause an outbreak if *Anopheles* mosquitoes were present, the Australian government launched an investigation in 1916 to determine the existence and possible extent of these mosquitoes in irrigation areas in the Murray River system, which had been earmarked for soldier settlement.[21]

The Murray River system, which included the Murrumbidgee River, was a site of significant intensification and expansion of agriculture in the first half of the twentieth century.[22] A range of political and cultural ideas and events motivated this expansion. At the turn of the twentieth century these

INFECTING | 79

included long-held ideologies of closer settlement for more intensive production, particularly of food; the political and economic fallout of what is now known as the Federation Drought (lasting from approximately 1895 to 1901); and the new legal framework of the national constitution following federation in 1901, which facilitated cooperation between states for river engineering along the shared waterways of the Murray system and in turn helped pressure NSW into developing irrigation along the Murrumbidgee, over which it had jurisdiction.[23] Soldier settlement in the irrigation areas was seen to support these goals, but the malaria parasites and mosquitoes significantly complicated these plans.

TROUBLING NATIONAL, TEMPERATE, AND BODILY BOUNDARIES

The Australian Quarantine Service appointed Frank H. Taylor, an entomologist at the Australian Institute of Tropical Medicine in Townsville, to carry out the survey of irrigation areas. In mid-October 1916, Taylor began visiting irrigation centers, starting in Kayabram in Victoria and, following the Murray River west, finishing in mid-January 1917 in Pompoota in South Australia. The distribution and redistribution of water was central to his study, along with mosquito preferences for water bodies in which to breed. Taylor's findings were summarized by the director of quarantine, John H. L. Cumpston: "*A. [Anopheles] (Nyssorhynchus) annulipes*—the Malaria Mosquito—is to be found right through the irrigation area, and, with the exceptions of Mildura and Renmark, wherever present they were found in considerable numbers."[24] Indeed, Taylor found *Anopheles* to be common in many areas he surveyed, and he thought the number of adults might be even higher in drier years, as many might have been killed by the heavy rains and floods that had swept through the river system just before he commenced his survey. While *Anopheles* were known to vectors in many places around the world, there were a number of species involved. This made knowing the details of this particular species, *A. annulipes*, and how it lived in these places even more important.

Taylor's findings soon provoked responses from the highest levels of government, as they held implications for national and state soldier settlement plans. In March 1917 prime minister William Hughes wrote to the premiers of NSW, Victoria, and South Australia that "the malaria carrying mosquito" was present "in such numbers as to render it exceedingly unwise to send affected soldiers to these districts." Consequently, medical checks of ex-servicemen were used to identify malaria sufferers through blood tests and oral testimony, removing them from consideration for settlement in the irrigation regions where *Anopheles* were known to be present. The Australian government viewed the introduction of malaria to these areas as likely enough that it became concerned about its plans for soldier settlement, but not certain enough to stop all soldier settlement there. There was still a risk that infected soldiers would slip through the screening process, but it was one the government seemed willing to take. Another possibility for malaria control was quininization, sometimes used as a prophylactic measure. However, this does not appear to have been proposed. While debates about the use of quinine versus mosquito control had been ongoing since the 1890s, the more immediate contexts of the Great War years, which had seen widespread confusion, controversy, and difficulty in organizing and enforcing quininization programs, perhaps influenced the government's decision not to pursue this.[25]

The government's initial response was to attempt to stop infected ex-servicemen from settling in the areas rather than undertaking a program of extensive drainage or mosquito eradication. Perhaps they regarded this as a quickly enforceable measure that could be enacted immediately and with relatively little expense, with both drainage and eradication programs requiring more planning and funds. In irrigation areas that both protruded into swamplands (also sometimes used for irrigation drainage and flood mitigation) and created new water bodies, significant water drainage was not possible. Controlling mosquitoes through oiling must have seemed, at best, difficult and in agricultural areas could have had consequences for pollinators (an issue raised in later investigations).

When linked to mosquitoes rather than climatic zones, malaria was no longer perceived by scientists and more widely as intrinsic to particular places and was seen, in many ways, as more easily eradicated. One newspaper columnist wrote in 1918, "Malaria was supposed to be a disease of tropical or sub-tropical climate until it was shown that drainage of the soil and the extermination of the mosquito, led to its abolition, though the climate remained as before."[26] Yet being freed, in a sense, from climatic zones (although temperature and other climatic conditions were still understood as influencing mosquito abundance and distribution as well as parasite development) and linked instead to the movements of mosquitoes and people, malaria was seen to be mobile in new ways that disturbed the boundaries between tropical and temperate.[27] Malariologists thus became concerned about the possible spread of malaria in Australia from northern into southern regions. R. W. Cilento, author of the first Australian book on malaria (published in 1924), described the possible spread of malaria into southern Australia via railway lines at the Australasia Medical Congress in 1923.[28] Starting in the late nineteenth century in Australia, the climates of temperate regions had been idealized, especially for white bodies, as promoting healthy, hard-working, civilized societies.[29] The involvement of Taylor, an expert in tropical medicine, in investigating mosquitoes in southeastern Australia and the possibility of malaria outbreaks in the Murray River system in itself subverted these boundaries, bringing the tropics to the temperate inland.[30]

Taylor's survey had largely been limited to areas on or close to the main channel of the Murray River and did not extend to the MIA, located on the Murrumbidgee River (see figure 3.1). However, the MIA was already receiving applications from ex-servicemen who were "malarial subjects." Faced with the likelihood that the *Anopheles* mosquito was in the area, and under threat from the Department of Public Health that the area could be closed as an "infected zone" if a malaria outbreak occurred, the WCIC introduced the same measures to exclude malaria sufferers from the MIA. The manager of a local returned soldiers' camp saw these checks as part of a holistic approach to assessing the health of retuned soldiers, including whether they carried venereal disease, adding that "the question of each man's medical fitness is important, from a Hygenic [sic] and Sanitary point of view; more especially such men as this scheme affects, and also to ascertain that each individual is capable of performing the duties of Farm life."[31]

Even without a survey, there were attempts to prevent mosquitoes from breeding in the MIA, more as a nuisance than a disease vector. The chief clerk of the WCIC offered to write a short summary for the local newspaper that drew "attention to the necessity of dealing with any stagnant waters about the farm if the mosquito pest, which is very annoying at [the local town of] Leeton, is to be combatted."[32] Local governments were required to prevent mosquito breeding in fulfillment of state laws, including the Local Government Act 1906, which specifically targeted mosquitoes. In the MIA, there was some action by the local government before Taylor's report, with one newspaper reporting in 1916, "The Local Government is out after the mosquitoes.... The ordinance [it has issued] provides penalties for persons who allow water to stagnate about their premises and thus breed mosquitoes."[33]

The NSW director general of public health proposed that, given the various risks, Taylor could undertake a similar survey of the MIA for the malaria mosquito. However, this was not done.[34] Perhaps the WCIC did not want to risk the potential for bad publicity if the "malaria mosquito," as *Anopheles* was called, was found. The MIA began operation in 1912 as a state-government-administered irrigation area intended to produce food from small, intensively cultivated farms, through horticulture, dairying, planting of alfalfa, and raising of pigs and fat lambs. The scheme was underpinned by broader ideologies of creating a "yeoman" class of small-scale farmer. Racially motivated ideas about desirable immigrants were reflected in the recruitment of farmers from the United States and Europe as well as elsewhere in Australia.[35]

The MIA had strong political backing but encountered problems early on, as many farms became waterlogged or affected by salinity. It soon teetered on failure, with one politician describing it as "a [political] hornets nest" in 1913.[36] A survey that revealed the presence of *Anopheles* mosquitoes at that time could have pushed the MIA into further scandal, just as soldier settlers presented an opportunity for reinvigoration.[37]

LINKING INSECTS AND AGRICULTURE, TOWN AND FARM

In 1919, one of the returned servicemen working at the Experiment Farm in Leeton developed malarial fever. The chief clerk for the WCIC in the area wrote to the head office that "this raises the question of whether a medical examination of returned men is altogether satisfactory as a method to detect those who are likely to be effected by malarial [sic]." That same month, J. H. Neilley, NSW director general of public health, pressed again for a mosquito survey of the area, writing, "Malarial mosquitoes have been found in areas not very remote from the Murrumbidgee Irrigation Area, and it is highly probable that they are present on the area." Neilley further noted that this would determine whether malaria sufferers could remain in the area over the summer. Neilley even offered for his staff to undertake the survey and asked that, if agreeable, the WCIC be "good enough to advise the earliest appearance of mosquitoes in the irrigation Area" so the survey could commence.[38]

Ultimately the WCIC agreed to a survey. Responding to the WCIC's notification in November 1919 that mosquitoes had appeared in the MIA, Drs. John Burton Cleland and Frederick Morgan, on behalf of the NSW Department of Public Health, made their way to Leeton.[39] Their two-day survey confirmed that the same species of *Anopheles* Taylor had found in other irrigation areas was also present in the MIA. Cleland wrote in his report that "one was caught near dusk on the verandah of the house at which we were staying." They found mosquito larvae in small pools rather than in open water. This included a large number of larvae in "the edge of a drainage channel containing little water," a small number in the overflow from an irrigation channel, and in hoof marks "on the edge of a drainage channel about 300 yards north of the Soldiers' Settlement near Yenda," with the aggregate larvae amounting to "a considerable number." Grazing livestock on the edge of channels was a common practice, and the seemingly small pools the animals created while walking were actually quite important for mosquito breeding. Cleland explained that *Anopheles* mosquitoes preferred these small pools because they were free from predators such as fish. He warned that unless measures were taken to reduce the number of mosquitoes, it was possible that there would be a number of cases of malaria in the area over the summer "should carriers sojourn in their neighbourhood." However, he thought an epidemic was unlikely.[40]

Presenting various options to the WCIC, Cleland advised that two lines of protection should be pursued: keeping malaria sufferers away and destroying malaria mosquitoes. Neither was sufficient on its own, he argued, as neither could be wholly achieved. The small pools should be oiled to asphyxiate the larvae, filled in, or joined to larger water bodies. He also advised further investigation of other possible breeding areas, such as swamps, water tanks, and the ends of irrigation channels. Residents could deal with possible breeding places on farms and around homes, while the WCIC could address the irrigation channels and swamps. He recommended a public health campaign, including instruction in schools, about mosquito destruction.[41]

Cleland found the malaria mosquitoes in out-of-the-way parts of farming systems, in close enough proximity to be a threat to humans. Humans were not an especially preferential food source for this species, which was just as likely to feed on other animals. Yet the favorable conditions created for them by irrigation settlements brought them into closer proximity to people and increased the likelihood that the female adults would bite and so transmit diseases to humans. In Australia and other settler lands, agriculture has largely been understood as bringing nature under the control of human activity.[42] Mosquitoes disrupted the sense of control over these places, creating uncertainty and vulnerability. Cleland further gave instruction on how to manage malaria sufferers. Those who were infected could be carriers for up to seven years, but he thought they were no longer dangerous after two or three years. He recommended continued screening of soldier settlers through blood tests and, if possible, that any sufferers detected in the area should be "removed or treated during the mosquito months of the year, under mosquito curtains." Cleland noted that identifying all sufferers would be difficult, as "many would not consult a medical man."[43]

The WCIC followed Cleland's advice. Despite the recent case of malaria, the resident commissioner claimed, screening had been "relatively successful in preventing the introduction of the disease." Yet with so much uncertainty surrounding the effectiveness of the screening techniques and the exposure of carriers to the mosquito, how close the area had come to an outbreak was anyone's guess, and the resident commissioner added, "Perhaps it has been pure luck." The resident commissioner, who was ultimately in charge of the WCIC's operations in the area, wrote that he would "encourage any practical method of destroying mosquitoes whether malarial or otherwise" by filling in and draining but held reservations about being able to successfully treat so many small pools of water over the expansive area. Further, before he supported kerosene oiling, he wanted some assurance that it would not injure any vegetation.[44]

Cleland arranged for another survey later in the summer by Dr. Eustace Ferguson, also from the Department of Public Health, who had gained firsthand experience of the disease in hospitals and then by running a field laboratory during the Great War.[45] In his three-and-a-half-day visit, Ferguson similarly found that *Anopheles* mosquitoes were fairly common in the area, mostly breeding in stagnant water and small pools rather than in main water channels or large drains where predators were present. He did not find mosquitoes generally very numerous and thought it "questionable whether the amount of breeding going on is sufficient to justify any widespread campaign of destruction." Yet he advised supervision of possible breeding areas and "taking steps to stop it," especially given that "the present year was not considered a bad year," and mosquito numbers might increase in later years. He recommended that a local WCIC agricultural entomologist, Keith C. McKeown, be entrusted with this surveillance work, as he had both the training and the interest.[46]

A few months later McKeown submitted a report stating that mosquitoes had been more numerous in previous years (1917–19) when there had been more rainfall. Yet even with a reduced number of adults, he noted, "in all parts of the Area so investigated Anopheline larvae were found," with most "in the vicinity of the town and human habitations." The breeding places ranged from a small pool formed by a leaky water main in Leeton "to the drainage channels, or from waste irrigation water, to pools in the holes formed by hoofprints of horses and cattle." The drainage channels, filled with stagnant water and vegetation, were especially appealing to mosquitoes because they provided good cover for adults. McKeown viewed the control of the mosquitoes as possibly much simpler than first assumed. Mosquitoes' preference for shallow pools of water and weedy drainage channels meant that eradication could be quite easy. However, he recommended that a more thorough survey first be undertaken. Since Ferguson had not regarded a malaria outbreak as an imminent threat, and given the seeming success of screening ex-servicemen, the resident commissioner decided that "no further action was necessary at present" and refused a further survey from the Department of Public Health the following summer.[47]

Modernist projects like the MIA promoted grand views of landscapes, filled with mighty dams and sparkling irrigation water, often seen as fulfilling notions of human mastery over nature. These mosquitoes forced a new understanding of water in the landscape, one filled with tiny water pools, stagnant drainage channels, and buzzing insects that came to be seen as lurking threats. The mosquitoes were in irrigation infrastructure such as drainage channels and in the hoofprints of farm animals, but they were just as likely to be in towns and along roads. Mosquitoes had not only transgressed the borderlands of nature and culture represented in the agricultural infrastructure, but they had crossed into the human heartlands of towns. Mosquitoes thus implicated people differently in the landscape, where intimate ecologies further blurred the perceived borders between nature and culture. These investigations thus created new imagined ecologies.

CIVILIZED TOTAL WAR

In the late 1920s, a new entrant into the imagined ecologies of the MIA significantly altered the water landscapes of the region, reigniting fears of a malaria outbreak. Following early difficulties experienced by farmers in producing planned crops for the MIA, rice became one of a number of agricultural crops under trial by farmers and the state government to replace unsuccessful products.[48] After initial trials of Japanese varieties imported by migrant Isaburo Takasuka and then successful trials of Californian varieties between 1920 and 1923, rice moved into commercial production, with the first such crop planted in 1924–25.[49] This was so successful that the WCIC encouraged farmers to grow rice and made water cheaply available.[50] Over the next decade rice growing expanded. By 1933–34, rice was farmed across about 20,500 acres, yielding 40,290 tons.[51]

Rice used flood irrigation and, despite its relatively high water usage, was deemed valuable by farmers and the state government because it seemed to suit the clay soils that underlay parts of the MIA that had previously been seen as unproductive. The clay soils held water well and were dense enough to prevent water tables from rising, thereby stopping potential seepage and salinization.[52] Rice cultivation also created suitable conditions for mixed farming, as the water held in the soil following a rice crop could be used to grow other crops or fodder for livestock.[53] Rice growing created new water landscapes, not only through the use of flood irrigation in paddies but also through the diversion of water into the MIA and the drainage of relatively large quantities of waste water from paddies into nearby swamps. Yet rice was grown during the summer months, when mosquitoes were most active.

Rice grown in the MIA went toward supplying a small domestic market, previously met by imports from Burma, India, Japan, and Java (approximately 20,000–25,000 tons of rice annually).[54] A newspaper article published in the *Sydney Morning Herald* in 1932 on a public speech by M. J. Gleeson, a member of the Rice Marketing Board (established in 1928), to the New Health Society shows how rice from the MIA was marketed against imported rice: "Prior to 1928 the rice consumed in Australia was grown in the East, harvested by natives, and marketed under the most unhygienic conditions. The rice for local requirements to-day, however is handled by machinery right from the field to the consumer."[55] Ideas about the vulnerability of white bodies to tropical diseases as well as modern technology were linked to the marketing and marketability of Australian rice for white domestic consumers.[56] Rice cultivation, so closely associated in Australia with the tropics and Asia, was thus creatively incorporated into the racial ideologies that underpinned the establishment of the MIA, with hygiene and freedom from disease at the core of the crop's marking.

In 1928, the Australian Council for Scientific and Industrial Research (CSIR) asked Ronald Hamlyn-Harris, city entomologist for Brisbane, "to investigate certain matters with reference to mosquito breeding in rice fields." McKeown, who had undertaken occasional work on the subject, was keen for further investigations because, in his view, "the mosquito problem is becoming increasingly serious," a situation made more acute by a recent outbreak of dengue fever—a mosquito-borne disease—in the area. The spread of a mysterious illness in nearby Narranderra in 1928 may have further added to fears of mosquito-borne diseases.[57] Managing mosquitoes on rice fields, McKeown wrote, "offers considerable difficulty and any work done in this direction will be of value." Unable to be filled in or permanently drained, the rice fields presented a significant challenge.[58] Hamlyn-Harris was formerly the director of the Queensland Museum. He had made eliminating mosquito-borne diseases such as dengue fever from Brisbane a key goal in his new role of city entomologist, the only one of its type in Australia.[59] He was part of a network of scientists that envisioned the emergence of a new, white "Australian race" adapted to local climates and environments.[60] These ideas were linked to ideologies of racial hierarchy, often couched in terms of standards of civilization and intelligence, which he brought to bear on his public campaigns to eliminate mosquitoes.

In January 1929, following his seven-week inspection of the MIA, Hamlyn-Harris delivered a public lecture in Leeton. He began, "The importance of mosquitoes in the world to-day is such that the civilized world undoubtedly gauges the intelligence of a people by the attitude they adopt in regard to mosquitoes. This is not my evidence but is the truth and you will find that the most civilized people have determined that the mosquito must go." He continued, "You can quite understand that it is that at last in 1929 the world has determined that to regard the elimination of mosquitoes anywhere as an impossibility is a sign of a lack of intelligence."[61] In this regard, it mattered little whether the mosquitoes in the MIA transmitted malaria or not, as this was a kind of "cleansing" of mosquitoes or "total war" that supported the fulfillment of ideology. This approach further reinforced a modernist vision of civilized control over entire landscapes, also evident elsewhere in mosquito eradication programs at this time and indeed earlier in the twentieth century.[62] This view of eradication, like extensive drainage, can be seen as an expression of totalizing hygiene that was simultaneously a colonizing and racialized process, framed as both dominating environments and marking civilizational progress.[63] Here mosquitoes were cast as a racialized criminal class that needed to be suppressed or "cleansed" from the environment and civilized society. Hamlyn-Harris's lecture and responses to questions afterward were infused with a language of hygiene where mosquitoes were portrayed as being literally dirty, as they bred in "dirty" drainage channels and tanks. Cleanliness and purity were seen in terms of civilizational superiority. His words resonated with the overarching ideologies of race and the civilizing influence of small-scale farming that had been central to the MIA's establishment.[64]

In official reports and the press, Hamlyn-Harris presented dire views on mosquitoes in the MIA. Reporting on a press interview with him, the Sydney *Daily Telegraph* noted that "mankind may be driven from the Murrumbidgee Irrigation Area by a plague of mosquitoes, and that the fertile area may become an uninhabitable waste." The paper quoted Hamlyn-Harris as stating, "Unless steps are taken to control the pests in the area . . . the mosquitoes will become so tremendously plentiful that people will not be able to go on living there. . . . There are days when ricegrowers will not go into the fields. I have not seen so many mosquitoes in any part of the world and their numbers are increasing."[65] Hamlyn-Harris's official report focused broadly on mosquitoes on the MIA, but it included specific information about *Anopheles* there. He wrote, "The rumours of a mosquito infestation in the Murrumbidgee Irrigation Area are more than justified" and "If they go on increasing at the present rate, settlers will find life on the Area a difficult problem." He derided the opinions of locals as "not sufficiently enlightened" and "loose talk," opining that "independent investigations" were therefore necessary. The swamps, which were used for irrigation drainage, had received "universal condemnation," but Hamlyn-Harris found "little justification" for this, although he added that they might be more favored by mosquitoes later in the summer. He suggested that the rice fields were blamed for more mosquitoes than they actually produced and that large quantities of harlequin flies, which bred and lived near the rice fields, were "invariably taken for mosquitoes." He found the actual incidence of mosquitoes breeding in rice fields to be "surprisingly small." As in previous reports, he rather found mosquitoes breeding mostly in drainage channels and small pools of water with "rotting vegetation," but also noted differences among the seven species he identified, with one species "showing an interesting tendency towards domesticity for on several occasions it was taken breeding in household receptacles."[66]

The section on malaria came toward the end of Hamlyn-Harris's report. He wrote, "The relative immunity from malaria in the presence of such large numbers of Anophelines bred in the Area is naturally noticeable." He argued that more research was needed "as to whether *Anopheles annulipes* is really the main vector of the disease in Australia," as "direct evidence is lacking." "It seems strange," he wrote, "that Australia should be the one country where this point has not been conclusively determined." This information was made more pressing given the "overwhelming" numbers of these mosquitoes in Australia. While the proximity of soldier settlers and *Anopheles* had "naturally" created concern, he thought use of "prophylactic measures" like quinine by returned soldiers would be common, so that "the mosquito itself is thereby rendered ineffective." Ultimately, his report cast "some doubt . . . on *Anopheles annulipes* being the most suitable vector of the disease in Australia and investigations to prove its capabilities and the conditions of transmission would be welcomed by those interested, and the scientific world at large." He wrote that though "malaria is essentially a household disease, transmission being only possible when certain ideal conditions exist, just what those conditions are it is difficult to determine." Yet there were a number of key factors—"any one of which may be an important factor for success or failure"—that he argued should be more widely known: "Not every malarial patient is capable of transmitting the disease. . . . A large number of infected Anophelines is necessary. . . . To become infective, Anophelines must take infected blood more than once. . . . Not every Anopheline is necessarily physiologically capable of becoming infective. . . . Temperature and humidity play a very important part as factors in the transmission of the disease." Despite these specific conditions being necessary for malaria transmission to occur, and his doubts about the effectiveness of *A. annulipes* as a vector of malaria, Hamlyn-Harris ended his report with a call for more "anti-mosquito" "publicity and propaganda" to remedy the "ignorance" of residents: "Just because so little is known about the subject the task of controlling mosquitoes seems beyond human possibilities." His recommendations for controlling mosquitoes included intermittent flooding of rice crops, use of oil-saturated sawdust, encouragement of larval destructors such as water bugs and beetles, and introduction of larvivorous fish into the MIA. Following his research in the area, local citizens attempted to form a mosquito elimination committee, but this was not supported by the WCIC, which regarded mosquitoes as not "sufficiently bad to induce people to go to any expense."[67]

The executive management of the CSIR was "doubtful about the wisdom of publishing" the report for general readership and saw more research as necessary, as Hamlyn-Harris had also indicated in his report, before conclusive statements could be made about mosquitoes in the area. Members of the WCIC administration, however, viewed publishing the report as the best way forward, perhaps to counter any negative publicity created by newspaper coverage of Hamlyn-Harris's investigation or to satisfy local calls for mosquito control. Hamlyn-Harris's report was not published, but perhaps as a compromise, the director general of public health gave a statement to the press that the chance of a malaria epidemic on the MIA was "remote."[68]

The main aspect of Hamlyn-Harris's report that was given further attention by the WCIC was his recommendation to introduce larvivorous fish, specifically a native species of hardyhead (*Craterocephalus fluviatilis*), into the irrigation channels and rice paddies.[69] Biological control of mosquitoes was a key subject of a paper Hamlyn-Harris later presented to the Royal Society of Queensland. While the main geographical focus was that state, he expanded his recommendations, stating that "this fish cannot be too highly recommended for mosquito work in the MIA," and suggested that breeding programs for the fish commence so as to build up stock for introduction to the area.[70] Yet the state director of fisheries held a different view, writing to the WCIC that enough of the native hardyheads were already in the reservoirs and rivers, hypothesizing that "many must find their way into the main irrigation channels," concluding that "there is no necessity to adopt Dr. Harris' suggestion." The WCIC does not seem to have further pursued the possibility due to expense and because, in the words of the manager of the local Mirrool Irrigation Areas, "it being felt that the mosquito menace was not at present sufficiently serious to warrant such action."[71]

MOSQUITOES AT HOME

During WWII, the possibility of another outbreak was raised in the WCIC. The manager broached the issue in 1942 as farms were set aside for soldier settlement, stating that the *Anopheles* had increased in numbers since returned soldiers were settled in the area following the last war, possibly because of the expansion of rice growing. The area under rice cultivation along the Murrumbidgee did not significantly change from the mid-1930s to 1941. In 1942, however, rice cultivation expanded by more than ten thousand acres in the MIA in order to supply the grain as food aid to parts of Asia where Japanese invasion and occupation had broken down established rice farming and trade, as food and resources were redirected by the Japanese and local, subsistence rice fields were taken over for other purposes.[72] In the MIA, the number of farms growing rice increased at the same time as the limit on acreage under rice cultivation on each farm increased temporarily. The state government also expanded rice cultivation to Murray valley irrigation districts such as Wakool.[73]

INFECTING | 91

For the MIA manager, greater amounts of water in the landscape seemed to have significantly increased the risk of a malaria outbreak ignited by returning soldiers. He also thought that "the Areas were probably lucky that malaria did not follow returned soldiers home after the last War." Another investigation was warranted, he reasoned, as "doubtless the knowledge of malaria has increased very materially." The manager recommended that infected soldiers again be quarantined from the area. The Department of Public Health was uncertain about how seriously to take the possibility of introducing malaria into the MIA. While the department recognized the risk of returned soldiers introducing malaria to Australia and worked with the Repatriation Department to treat all soldiers until they were "cured, or at least, are no longer infective," the director general wrote to the WCIC that "the fact remains that, after the Great War [of] 1914–18, a large number of soldiers who had been infected by malaria returned to NSW—and probably many of them settled in the Irrigation Areas—but they did not act as reservoirs of the disease." Yet there was always the possibility—that grain of uncertainty that "cannot be absolutely eliminated"—that this time returned soldiers would cause an outbreak.[74]

Under the specter of a growing number of cases of dengue fever in NSW, the state government launched public awareness campaigns that employed the imagery and language of war. Disease-carrying mosquitoes were another enemy that needed to be destroyed. In this context, in 1943 the manager of the MIA again voiced concerns about malaria in the WCIC, writing that "we are in a dangerous position not only in the State as a whole but on these Irrigation Areas in particular." He noted that there had been a recent case of malaria in Bega, on the NSW coast, where a child who had never left the district contracted the disease. As increased rice production had increased mosquito numbers, the risk had also risen. Further, there was a patient in the Leeton hospital suffering from malaria: "He is in an isolation ward, but that is not sufficient to prevent a mosquito from passing the trouble on if—and this is a big question about which nobody seems certain—our anopheles mosquito is able to transmit malaria. Surely it is time somebody found that out for certain!" Doubt about the ability of this *Anopheles* to transmit malaria was crucial to the imagined ecologies of the MIA.[75]

In 1944 the Department of Public Health advised the WCIC that Frank Taylor would undertake a new survey of *Anopheles* mosquitoes in parts of the state, including the MIA and Wakool areas, where rice growing had expanded during the war. Taylor completed his survey in late April 1945,

FIG 3.3 The imagery and language of war infused mosquito eradication efforts in World War II. This Australian government poster targeted mosquitoes as vectors of dengue fever, and therefore "as dangerous as an enemy bomber." "As Dangerous as an Enemy Bomber: Mosquitoes Carrying Dengue Fever Can Cripple an Army," NSW Dept. of Public Health, c. 1940–42. Image courtesy of the National Archives of Australia, NAA MP742/1, 211/6/1057.

finding abundant *Anopheles* breeding areas in both the irrigation "supply and subsidiary" channels, especially where they were "overgrown with vegetation." Like Harris, he recommended these be kept "clean." The swamps in the irrigation area also posed a problem, and he advised that they be "filled in" as soon as possible. Yet, he warned, attempts to reduce mosquito numbers in this way could also lead to reduced numbers of small flies that bred in similar places, which were important for pollination of fruit trees.[76]

Since the other reports had been written, a new technology in mosquito control had become available: DDT-based sprays. Developed during World

War II to protect military forces from malaria mosquitoes, the insecticide was used around the world in the postwar and Cold War years for a range of applications, notably as an agricultural and household insecticide and for spraying water bodies to kill mosquitoes.[77] Taylor promoted the use of a DDT spray in homes to keep adult mosquitoes away. He wrote, "I have used this spray in my home with very marked effect, simply by spraying the window frames, windows, and the electric light cord and shade." Having breathed in the spray mist for several minutes himself, he felt confident that "there is no risk in using this spray." The benefit of the spray was its residual effect: a single spraying could keep mosquitoes away for two to three months before another application was needed. Taylor explained, "If this spray is used as suggested it will not matter how many *Anopheles* come into dwellings because *they would not leave the house alive*."[78] Mosquitoes threatened the safety of the home, but using DDT could help humans reassert control over "their" domestic spaces, thereby maintaining a boundary from within, between the home and the lurking threat of these nondomesticated nonhumans.[79]

Like Hamlyn-Harris, Taylor emphasized that a number of circumstances needed to align for malaria to be transmitted and that the risk in an area where it was not already endemic was actually much less than might be thought:

> Only a small percentage of such cases have gametocytes in their blood, only a percentage of the mosquitoes which bite these cases become infective. Of those that become infective only some will survive to bite another man. Those that survive in that way will be limited in the range of their dispersal. Those living within a few hundred yards of a case of malaria will run the great risk of infection. It will be necessary for a number of infective mosquitoes to bite an individual for that individual to get malaria. Ross stipulated that a minimum of ten (10) Anopheles were necessary to bite a person for that person to contract malaria.[80]

Further, based on the fact that "only isolated cases" of malaria had occurred in NSW, he asserted that the "*Anopheles annulipes* is definitely a very poor transmitter of malaria." For the mosquito to cause an epidemic, he wrote, "it would be necessary for very large numbers of that mosquito to bite the inhabitants of the infested area; probably the number would run into hundreds of mosquitoes per person."[81]

In fact, entomologist and parasitologist Frederick Hugh Sherston Roberts had just published the results of a laboratory experiment on the effectiveness of *A. annulipes* as a vector of malaria parasites. While so far "observational evidence" had shown the mosquito to "at best be a poor natural vector," his experiments built on some previous tests of the mosquitoes to show that they were a very capable vector under laboratory conditions. He concluded, "Any deficiencies which *An. annulipes* may show as a vector of malaria in nature are not due to the defects of its hospitality to the parasites." He argued that other factors might be "more important," such as "abundance, feeding habits, and other biological characteristics of the insect" in obtaining "a reliable pointer to the degree of respect with which it must be treated." Yet Roberts cautioned "that until this additional evidence is obtained, it would be unwise to concentrate large numbers of gametocyte carriers in places where *An. annulipes* is abundant, unless adequate protective and control measures are carried out."[82]

Despite Taylor's view that the mosquitoes were poor transmitters of malaria, he also took a precautionary approach, recommending that the interiors of all soldiers' homes in the MIA be regularly sprayed with insecticides containing 5 percent DDT. If this was done, he thought, there would be "absolutely no risk from malaria in any part of the irrigation areas."[83] The WCIC and NSW Department of Public Health investigated the possibility of enforcing DDT spraying in homes, but ultimately the Department of Public Health decided that the chances of a malaria outbreak did not warrant "any compulsion, at present, in the use of such sprays." Yet when the WCIC drafted a summary of Taylor's report to release to the local press, the department advised that they should "include mention of the use of D.D.T. spraying for the elimination of mosquitoes in dwellings." This was carried out with an article published in a local newspaper advising residents to use sprays in homes that contained "not less than 4 per cent D.D.T.," which were available from local stores.[84]

In the 1950s new kinds of imagined ecologies took hold, and human nature was again remade by dynamic interspecies relationships. While researchers in the early twentieth century had sought to control or eradicate *Anopheles annulipes*, in the 1950s they sought to harness these mosquitoes to spread myxomatosis among rabbit populations, which was introduced as a way of controlling these widely derided pests. From the 1950s attention also moved to bird reservoirs for various and newly isolated mosquito-borne diseases, focusing on avian movements between the Murray River basin,

Papua New Guinea, and parts of Asia.[85] Scientists currently view these mosquitoes as circumstantially linked to a small number of cases of locally acquired malaria in southern Australia.[86] Scientists currently understand *A. annulipes* (like many mosquito species) to be a species complex rather than a single species, with some uncertainty over the feeding preferences and vector abilities of different groups, including their ability to transmit malaria.[87] However, the potential for these mosquitoes to cause an outbreak still influences public health operations and research. In 2008 scientists were asked to address local concerns in some areas in NSW that increased numbers of refugees from endemic malarial regions would cause an outbreak and recommended screening of this group due to the prevalence of *A. annulipes* in the state.[88]

In many ways this is a history of how some of these understandings were formed (with little prior research having taken place), written with a commitment to the view that knowledges are always situated, partial, and emergent.[89] Underpinned by powerful political and social goals, watery farming areas like the MIA could not be drained nor easily oiled, presenting new medical, environmental, and ideological challenges. This emergent, situated scientific knowledge combined with public health concerns influenced new imagined ecologies that shaped new kinds of relationships between humans and nonhumans. These imagined ecologies reveal the ways understandings of environment, animals, and plants are world-making and how those understandings and relationships have been coconstituted with nonhuman organisms in dynamic and historically significant ways, with consequences for all those involved. Changing understandings of mosquitoes and irrigation systems in the MIA in the first half of the twentieth century, together with two world wars, further reveal some of the shifting biocultural terrains of the borders between temperate and tropical, civilized and uncivilized, the home and nonhumans.

FOUR

CROSSING

Wildlife in Agriculture

AS WE FLEW ACROSS THE MURRUMBIDGEE IRRIGATION AREA, THE water glinted in the sunlight. I could make out irrigation channels and pools of water across the landscape that had not yet dried after the recent floods. Ducks and other birds gathered on these, occasionally taking flight across the checkerboard farms. In the summer months these farms intermittently filled with water to grow rice, supplied by the irrigation channels and by large dams upstream on the Murrumbidgee River and its headwaters, such as Burrinjuck Dam and those of the Snowy Mountains Hydro-Electric Scheme. The farms were now mostly brown stubble, as the rice had been harvested and a winter crop had not yet been planted. I was in a light aircraft piloted by a rice farmer who had generously offered to take me on a tour of the area from the air. His farm was located in the heart of the MIA, directly between the two wetlands of Fivebough and Tuckerbil Swamps. He had told me that one of the things he loved about the location of his farm was all the birds that came to it as they moved between the wetlands.[1] First, we flew west over Tuckerbil Swamp, a shallow, muddy wetland so full from floodwater that it merged into the bays of an adjoining rice farm. We saw more ducks, as well as black swans, water hens, and white ibis. After about ten minutes, we turned sharply and started heading east to Fivebough

FIG 4.1 Rectangular bays of rice farms adjoin the edge of Tuckerbil Swamp. Floodwater in the swamp has filled some of the bays, blurring the boundaries between wildlife and agricultural areas. Birds dot the watery landscape, moving easily between these areas. During the rice-growing season, farmers flood the bays, and birds such as ducks are drawn to this water. This has historically caused conflict and controversy involving ducks, farmers, government scientists, and hunters. Photo: Emily O'Gorman, 2012.

Swamp. This presented a somewhat different sight, with stretches of richly vegetated edges as well as some mudflats, and a wooden platform for viewing the many birds both on the water and in flight. From the air we could clearly see holding bays for treated sewage in the swamp. Many birds dotted the surface there too. In flight the birds moved easily across the sharp lines of the farm boundaries, now often heading toward the swollen swamps. But I knew that at other times the birds could move easily in the opposite direction, onto the farms, as my pilot had told me. When the bays were full and the rice was growing, many birds followed the water into the paddies.

This chapter stays with the changing water landscapes of the MIA in NSW in order to focus on the shifting interface between agriculture and wildlife in the twentieth century, a period of agricultural intensification and expansion supported by large dams and irrigation networks. This topic is important in a history of wetlands, as this kind of agricultural expansion has remade wetlands throughout the Murray-Darling Basin and around the world, and it has often done so over relatively short periods of time. Indeed, today one of the most prevalent political issues in the basin is the conflict pitting the water needs of farms and agricultural industries established in this period against the ramifying ecological degradation from these water extractions, with consequences for downstream farmers and Aboriginal cultures and livelihoods.[2] In the twentieth century, huge engineering schemes moved large quantities of water to new places, not only diverting water from existing wetlands but also creating new water bodies. Wetlands further became enrolled in new agricultural enterprises in a variety of ways, including being utilized for cattle grazing, flooded, and drained. Different organisms navigated and cocreated these changing landscapes in various ways. In these shifting environments, what has counted as a wetland for whom has had significant consequences.

This chapter considers more closely the MIA's changing relationship with Fivebough and Tuckerbil Swamps, as well as the movement of birds back and forth between different environments. It examines who has had a stake in establishing and maintaining particular boundaries between areas for agriculture and for wildlife and some of the repercussions of these demarcations as wildlife have sought to move within and across these landscapes. As other scholars have noted, animals have often disrupted these boundaries with varied consequences.[3] While many birds and other animals have engaged in such movements in this region, native ducks have taken on particular significance, disrupting boundaries as they crossed into the rice fields of the MIA. Key to this particular disruption is that these ducks are nomadic rather than migratory—that is, they follow water rather than seasons, as ducks in North America and Europe do.[4]

Ducks started visiting flooded rice paddies in the MIA shortly after farmers began to grow rice there in the 1920s. Many farmers (but not all) saw ducks as pests as they muddied the water, pushed down crops, otherwise hindered the growth of rice, and were widely perceived to eat rice. Native species of ducks have continued to visit the rice fields ever since. In dry years the paddies have presented a watery refuge, as they were often wet even when many wetlands were not, and in wet years they have provided an extension of a generally wet landscape. Ducks have remained controversial figures in the region, occupying a complex and multifaceted space for managers, biologists, farmers, and hunters. These groups have variously portrayed them as wild or native and so belonging, or as agriculturally disruptive and invasive. Some farmers and hunters have viewed them as a pest to be controlled, while others have valued them as game birds or for their role in eating invertebrates that damage crops. Ducks have therefore brought together many different interests and long historical legacies related to issues such as maintaining populations for hunting, the conservation of wildlife, and agricultural economics. In the early to mid-twentieth century, these histories and questions all came to bear on the controversies over whether ducks damaged rice crops and how they should be treated in these areas. The involvement of two government biologists—James Kinghorn and later Harry Frith—provides a lens onto these controversies as well as onto changing ideas of pests and conservation in this period.

The ducks invite us into a different perspective on wetlands—and more-than-human histories—because they treat rice paddies as wetlands but are not welcome on many farms. They show us a different kind of storied landscape in which what has counted as a wetland for them in these changing places has been driven by different needs and interests.[5] Since rice was first grown commercially under extensive irrigation in the MIA in the mid-1920s, water diversion for flooded rice paddy irrigation has meant that these agricultural landscapes have sometimes replaced established wetland habitat for animals such as ducks. The expansion of agriculture changed wetland environments and so reconfigured possibilities for animals and plants, with mixed outcomes for all those involved.

SWAMPS IN AN IRRIGATION LANDSCAPE

In 2002, the Ramsar Convention jointly listed Fivebough and Tuckerbil Swamps as internationally important wetlands. The swamps are approximately ten kilometers apart but are hydrologically connected by a floodway. Both have a high diversity of waterbirds, some of which are endangered, like Australasian bitterns (*Botaurus poiciloptilus*).[6] The swamps are located in the MIA, an important center of rice growing in Australia.[7] They have been deeply shaped by irrigated agriculture since the establishment of the MIA in 1912, constructed by the NSW state government in the context of increasing popular support for irrigation.[8]

This part of the Murrumbidgee River system is the Country of Wiradjuri Aboriginal people, who have long used Tuckerbil Swamp in particular for hunting and fishing. There are important Wiradjuri sites around these wetlands, including an ancestral burial ground near Tuckerbil Swamp.[9] Following the often violent frontier conflicts between mostly Anglo-Celtic colonists and Wiradjuri, which reached their height in the late 1830s, many Wiradjuri took up employment on the pastoral properties that dominated the region from the 1840s, mostly as a means to stay on Country. Fivebough and Tuckerbil Swamps were part of a grazing property, and it is likely that these wetlands would have been used for grazing in the nineteenth and early twentieth centuries.[10] The establishment of the MIA created further upheaval for Wiradjuri, and in the first two decades following its creation, more Wiradjuri moved into missions or to places outside these intensively cultivated regions.[11]

Both swamps had black box–lignum vegetation (a kind of wetland woodland scrub) when the MIA was constructed. Both were also ephemeral, meaning they would dry up and then flood during wet periods. In 1920, ornithologist Samuel White argued that the importance of these swamps to birds suggested they should be made into reserves. The water regimes and vegetation of the swamps changed substantially in the 1930s. A damaged drainage pipe flooded Tuckerbil Swamp for a prolonged period, significantly altering the water regimes. Fivebough and Tuckerbil Swamps, along with nearby Barren Box Swamp, were also used for irrigation water drainage and storage, and this usage seems to have coincided with the expansion of rice growing in the areas in the late 1920s and 1930s. Farmers grew rice through flood irrigation, and it seems likely that the flooded rice bays were drained into the swamps, where water would then evaporate.[12]

Some people saw these changes in the swamps' water regimes as having positive effects, particularly on bird life. In 1939 Keith McKeown, the NSW government economic entomologist in the region, wrote an extensive article in the *Sydney Morning Herald* about the bird life in the area, which he argued had increased with the greater amounts of water brought by irrigation. His article also publicized the Royal Australian Ornithological Union's upcoming annual congress to be held in the MIA. Of the two swamps, he wrote, "With the coming of irrigation, these low-lying areas became semi-permanent sheets of water. The waste water which flowed into and filled them was at first limited.... Today, with the increase in irrigated acreage the swamps ... have become permanent ... and have become a fixed home for many aquatic birds." Whereas birds had previously left when the swamps dried up, they were now there year-round, and he argued that the swamps should be declared "sanctuaries."[13]

The changed water regimes, however, began to alter the vegetation, and many of the black box trees died.[14] The vegetation was gradually replaced by water grasses, like Cumbungi and water couch. In 1957, biologist Harry Frith wrote that these now "large permanent swamps" were "presumably ... relatively sterile"—in terms of food for birds—as a result of these vegetation changes.[15] Along with changes in water regimes, this alteration in the condition of the swamps was possibly also because water that was drained into them from rice farms contained potent pesticides and herbicides, which were used extensively in the postwar period, including DDT and 1080.[16] In the 1960s and 1970s both swamps were used for grazing. As a result, Fivebough turned into a mudflat, which attracted shore birds.[17] Since the 1930s, Fivebough was also used to hold floodwater as part of mitigation measures for farms and the nearby town of Leeton (established as part of the MIA). In the 1970s Fivebough became part of sewage treatment for Leeton, and from the 1980s treated sewage was released into the swamp for disposal by evaporation.[18]

These changes created new conditions for a range of species. For example, shorebirds were attracted to the mudflats created by grazing, and other birds as well as frogs to the treated sewage bays. Indeed, the Ramsar listing for these sites highlights the importance of past and contemporary uses for the biodiversity of the area: "Both wetlands operate as important waterbird habitat and refuge within an agricultural landscape and in fact gain some of these habitat values from human uses of the areas such as grazing, flood mitigation and sewage treatment."[19] Other species have not done as well, including black box–lignum vegetation and those reliant on it.

Fivebough and Tuckerbil Swamps were the first wetlands in Australia to be listed in the Ramsar Convention with full knowledge that they had undergone radical transformations since, and in many ways due to, British colonization.[20] This was not an effort to save remnant ecologies, as other Ramsar listings had been; it was an effort to prevent further development. The listing was controversial at the time and remains so today. It followed a local proposal in the 1990s to turn Fivebough Swamp into a permanent recreational lake. Other members of the local community instead argued that the now intermittently flooded and dry swamp was valuable to birds, especially shorebirds attracted to the mudflats, and that it should rather become a tourist attraction for bird watching. They were supported by nongovernment organizations and lobby groups like the Inland Rivers Network and World Wildlife Fund.

Cath Webb and Jamie Pittock worked for these nongovernment organizations at the time and were involved in listing the swamps with the Ramsar Convention. They also happened to be married. They recalled that there was a backlash from other environment groups and academics because by listing the swamps in their current state they were in some sense accepting the impact of cattle grazing, use of the swamps for irrigation drainage, and so on. At the same time, Cath and Jamie aimed to support and build a rural constituency interested in conservation, and they recognized the many changes that wetlands in the Murray-Darling Basin had undergone since colonization and were now facing. Jamie led a program at the World Wildlife Fund that aimed to list a Ramsar wetland on each major tributary in the Murray-Daring Basin so as to maximize the benefits of environmental flows, a recent policy innovation intended to guarantee water for the environment and not just agricultural and other industry interests. However, Fivebough and Tuckerbil Swamps were not on the main channel of the Murrumbidgee River and presented very different prospects from what he had envisioned. Listing these swamps meant that the then-current uses would need to be continued to maintain the nominated values. Jamie noted:

> One of interesting things about Fivebough is that it was the first time in Australia that a site was deliberately listed knowing that it wasn't in a vaguely pre-European ecological state. . . . It had been entirely transformed. . . . Other places had been listed, but it wasn't done in a thoughtful and informed way. . . . This was a deliberate decision to list something that everyone knew would have to be managed with continual impact, and so one of the early dilemmas became how to graze it in a way that kept it open enough for the migratory shorebirds to keep coming and using the place as habitat even though it was now an internationally significant wetland, and it hadn't been common practice in managing conservation reserves in Australia before.[21]

Following the listing of Fivebough and Tuckerbil Swamps, local community and management groups undertook various environmental programs, often centered on Fivebough Swamp, including planting additional vegetation, establishing interpretive signs, and building viewing platforms and walkways. These swamps, especially Fivebough, have come to be managed for birds—particularly shorebirds—and bird watching, with potential consequences for other species.[22] Yet the behavior of some birds, particularly ducks, in this and the wider area has been a source of conflict between groups with a long and controversial history.

DUCKS IN RICE FIELDS

Some wetland species, including birds, move between the swamps and surrounding farms—for example, various native Australian ducks that have a history of visiting the flooded rice paddies of the MIA. Rice's semiaquatic characteristics require that paddies be flooded at particular times in the growth cycle to ensure high yields and control weeds. The plant's relatively high water requirements have led to heavy criticism of its cultivation in recent national debates.[23] In addition, there are a number of wetlands along the Murrumbidgee, such as the Lower Murrumbidgee floodplain—important sites for bird breeding and biological diversity more broadly—that have received reduced amounts of water due to upstream irrigation extractions.[24] The loss of wetland habitat in the last three decades has led to an overall decline in numbers of some duck species like grey teal (*Anas gracilis*) throughout eastern Australia. Yet the changing water landscapes of the Murrumbidgee and other river regions have also been inhabited by waterbirds in various ways. For instance, populations of Australian wood duck (maned goose, *Chenonetta jubata*, which is actually closer to geese than ducks) and Pacific black duck (*Anas superciliosa*) have increased in recent years, partly through their inhabitation of increasing numbers of farm dams.[25]

The populations and movements of these and most other species of native Australian ducks are considerably influenced by the wet and dry periods that characterize eastern Australia. These ducks flock to rivers and wetlands in flood, to feed, nest, and breed. The species of duck that visit the Murrumbidgee region have largely remained the same throughout the twentieth century. They include the grey teal, which are extremely nomadic and can fly large distances to flooded waterways and wetlands.[26] Other species include the Pacific black duck, the Australian wood duck, the pink-eared duck (*Malacorhynchus membranaceus*), the Australian shelduck (mountain duck, *Tadorna tadornoides*), and the plumed whistling duck (plumed-tree duck, *Dendrocygna arctuata*).[27] The population of each species significantly increases during large floods, as females hatch several broods in a few months, and crash during droughts, when breeding slows. While breeding can occur at any time of year, spring river rises in the Murrumbidgee and Murray from melting snow often lead ducks to breed there in these months.[28] Aboriginal groups along the Murrumbidgee River—Wiradjuri and Ngunawal—have traditionally hunted these ducks and eaten their eggs, particularly during floods.[29]

Throughout the twentieth century ducks in this region have been a source of conflict and cooperation among groups such as managers, conservationists, farmers, and hunters. These interactions have centered on whether the ducks damage rice crops and how they should be treated in these areas. Conflicts emerged in this period and have continued amid changing contexts, including the expansion of rice growing following WWII, and again in the 1960s. Biologists, and later ecologists, conducted a range of investigations, often at the request of farmers, to resolve whether ducks significantly damaged crops and to gain government support. In broad terms, as rice growing expanded, so too did concerns over duck damage. Aspects of these conflicts changed over this period, but perhaps most strikingly, the conflicts are still ongoing.

Ducks have been attracted to flooded rice fields both as drought refuges and as aquatic habitat during local floods. In the words of ecologists Alison Curtin and Richard Kingsford, "When rice is flooded, it creates a wetland."[30] However, the sharing of space and resources by agriculture and wildlife in Australia has rarely been a matter of easy coexistence, as many species, such as ducks, have been and continue to be hunted or targeted in other ways as agricultural pests.[31] In the late 1990s, managers put forward the irregular and large variations in duck populations, which change with droughts and floods, as a major reason for failing to find management solutions for sharing these landscapes.[32] This was in contrast to more effective management in other countries (albeit contested and highly politicized), such as the creation of wetland reserves in Californian rice-growing areas, which catered to the regular seasonal migrations of waterfowl.[33]

Ducks, like other "'pests' that insinuate themselves into agricultural systems," disrupted boundaries between wildlife and agriculture.[34] While some wildlife have been allowed or even encouraged on farms and infrastructure, these are often benign or in service to the farming project, such as insect-eating birds. Here, what farmers and others see as belonging on farms and in agricultural systems is firmly demarcated. This is clearest in the treatment of agricultural pests, through killing and poisoning, to increase yields or protect livestock. Like these pests, ducks "threaten[ed] the enactment of a firm border between nature and culture" by inhabiting agricultural systems and upsetting notions of human control.[35]

DUCKS AND THE ESTABLISHMENT OF RICE: KINGHORN

In 1932, James Roy Kinghorn, a zoologist at the Australian Museum in Sydney, published the results of his investigation into whether ducks damaged rice crops in the MIA. He began by explaining the background to this research, which focused on two rice seasons, the summers of 1926–27 and 1927–28. During these years widespread drought conditions had attracted "thousands of wild ducks and other waterfowl" to the region.[36] In the first of these seasons, some growers blamed their reduced crop on duck damage, arguing "that both the black duck and the grey teal ate the fresh sown seed and sprouting plants, puddled the bays (thus interfering with the young plants), and later destroyed the rice when it was in head."[37] The drought continued into the following rice season, and the ducks stayed. Fearing more damage, a group of farmers attempted "to have the names of several species removed from the list of protected birds" during an enforced closed season so that they could be hunted as pests. However, other farmers claimed that the ducks had done almost no damage and in fact helped their farming by eating weeds in the rice bays. Because of this "contradictory evidence, and as there was no information available as to the economic value of wild ducks in regard to rice cultivation," Kinghorn was "asked to investigate." He initially tried to undertake this research from Sydney, by examining the gizzards of seventeen "ducks forwarded from the Area," of four different species. However, he became wary of this proof, "as it was evident that some of the ducks, prior to being killed had been forcibly or purposefully fed with mature rice grain." Kinghorn therefore decided to make his own field investigations.[38] This was only the third season of a commercial rice crop in the MIA, and controversies around ducks in rice fields had already begun.

During the first commercial rice season, farmers had reported damage to rice crops to the Australian Museum, requesting a special open season on ducks.[39] The Australian Museum was established in the early nineteenth century as a natural history museum, and its primary purpose was the collection of animal specimens and taxonomic research. In the early twentieth century, parts of the museum were redirected toward the needs of agricultural departments, as at other research institutions in Australia. For instance, the museum undertook applied research into animal biology and behavior that related to agriculture, which drew on its strengths in zoology.[40] The museum recommended to the NSW chief secretary—who declared open seasons—that the open season currently in place in the region, from February to April, be maintained and not altered to better suit the times when rice was vulnerable, from approximately September to March.[41] This was possibly due to concerns that hunting across the summer months would coincide with the peak of the duck nesting season. Duck breeding seasons had traditionally been closed hunting seasons, so as to protect duck populations for game shooting. Another reason may have been some farmers' assertions that the birds ate insects that damaged rice crops.[42]

A growing controversy over whether the ducks acted as a pest that hindered the rice crop or helped by eating weeds sparked Kinghorn's investigations. Kinghorn visited the MIA three times during the 1927–28 rice season, at different stages of cultivation and plant growth. During this time, he talked to farmers, observed birds in rice fields, and hunted ducks in rice areas and nearby wetlands such as Fivebough Swamp for his own gizzard analyses (totaling seventeen ducks from four species). Kinghorn's investigations reveal his unfolding views about ducks and agriculture. They also show some of the views and practices of farmers and the ways ducks and other birds were inhabiting these new aquatic environments.[43]

Kinghorn found that grey teal, black duck, and wood duck ate rice seed at the time of sowing as well as green rice plants, but also barnyard grass weeds, "weed seed," and some insects. Pink-eared ducks, however, ate no rice and were instead valuable "as an insect destroyer." He also argued that waterbirds like the glossy ibis, cranes, herons, white ibis, and spoonbills were helpful to rice farmers as they ate insects and crayfish, the latter of which could undermine check banks, drains, and other earthen infrastructure. Further, Kinghorn found that ducks were not the only birds that potentially damaged rice crops. As the first flooding took place in the rice bays, he saw starlings and crows retreating before the advancing water, "collecting worms and insects which were wriggling hurriedly from the soil," "but at the same time there appeared to be no doubt that they were also eating the freshly sown rice seed." To Kinghorn it seemed that ducks had been blamed for the damage these birds, and others like galahs and sparrows, had caused, as some farmers considered ducks to be "the only bird" to damage rice.[44]

While ducks puddled and pulled out some rice, Kinghorn argued that much of the damage attributed to them was instead due to "faulty farming methods." He reached this conclusion through his interviews with farmers, where he found that "it was very noticeable that the growers who produced the best rice crops had little or nothing to say against ducks." Kinghorn claimed that bare sections in a rice crop were mostly due to rice that had been broadcast washing into depressions and then drowned as water pooled there when the rest of the bay was drained. Similarly, rice could drown if the bays were poorly graded. Broadcasting seed also seemed to produce a weaker young rice plant than drill sowing. Ducks therefore more easily puddled plants grown from broadcast seed, which were often blown onto the banks by wind.[45]

The main conclusion of Kinghorn's report was that overall "wild ducks" were "not a serious pest of rice crops." While ducks as well as other "native companions" could do some damage if they descended in large numbers during droughts, he found that it was "impossible to say" the extent of this damage, as there were many other factors, and critters, involved.[46] Kinghorn pointed to a complex situation emerging in rice areas, which was not just about ducks and rice but many other connections and relationships.

While we can perhaps recognize an ecological approach attentive to nested relationships in Kinghorn's study, he did not couch his research in these terms. Ecology was not then widely taken up as an applied science, and there is no evidence that Kinghorn took a particular interest in the emerging field.[47] Kinghorn's approach is perhaps better understood through changing approaches in field biology, and his research echoes other field studies undertaken around the world at the time. For instance, Herbert Stoddard's work on the decline of quail numbers in Georgia from 1924 to 1929, undertaken for the US Agricultural Department's Bureau of Biological Survey at the request of hunters, took a broadly comparable approach (albeit to a very different question), and Stoddard also claimed that a central factor involved farming practices that adversely affected the quail. Historian Thomas Dunlap has argued that Stoddard's research was not explicitly ecology, but it nevertheless "was marked by a new view of nature" that examined the complex interacting behaviors of humans and nonhumans.[48] Similarly, Kinghorn placed people in the landscape in a relatively new way by arguing for the importance of farming practices.

Kinghorn thought some farmers' belief that ducks were responsible for significant rice damage was mostly "the outcome of imagination and founded on hearsay."[49] Perhaps preempting a backlash from farmers, in his unpublished report (which appears to have been available to growers), Kinghorn wrote, "For the information of the growers, I would like to say that I went into this matter completely unbiased, and in the early days of the investigation was inclined to believe ducks were doing a lot of damage. As time advanced my opinion changed and I was eventually able to prove to my satisfaction that the alleged damage was exaggerated."[50] Yet enough farmers maintained that ducks did do significant damage that after further complaints in 1932, the NSW government allowed them to apply to shoot ducks, waterhens, and red-bills on their property from August to December.[51]

Farmers may have dismissed Kinghorn's findings as "top-down" science, which they often resented. Further, farmers may have disagreed with Kinghorn about what constituted "serious" or acceptable damage. For some farmers, any losses to ducks, or any animal seen to be a pest on their farm, could have been seen as unacceptable. In some cases, farmers' views that no ducks should be present on rice farms seem to have been underpinned by a dominating view of the landscape, in which farmers had ultimate control over what did and did not fit on their farms. That is, if farmers did not regard ducks as belonging on rice farms, then they would be forced off the farms or killed.[52]

NATIVE PESTS, THE ECONOMIC VALUE OF BIRDS, AND HUNTING

These debates about ducks in rice fields draw attention to some of the wider histories of how people have understood the relationships between native animals and agriculture, particularly competing economic views of some native animals as pests or as helpful to farming through their eating of weeds and other animal and insect pests. In 1880 in NSW, the government protected the killing of native and introduced fauna that colonists regarded as agricultural pests with the Pastures and Stock Protection Act. This act initiated a system of bounties on particular native and introduced animals that lasted for fifty years but did not include ducks or other birds.[53] From the mid-nineteenth century to the early twentieth century birds were largely included in legislation under game laws intended to conserve game populations by limiting overhunting, initially through a series of years of complete bans on hunting and then through the introduction of open seasons that aimed to protect birds during their breeding seasons. Many native ducks, including those that later visited rice fields, took on the role of similar species in European hunting activities and were included under these laws as game birds for sport.[54] The first of these acts, the Game Protection Act 1866, explicitly exempted two groups of people from hunting restrictions: government-approved collectors of natural history specimens and Aboriginal people.[55] These exceptions were carried forward in subsequent legislation.

Native and imported game came under different sets of rules in colonial legislation. For instance, native ducks were included in the Birds Protection Act 1881 under the general category "wild ducks of any species" within the broader category "native game," as opposed to "imported game" such as pheasants. In this period and throughout the twentieth century, colonists used the terms "wild" and "native" almost interchangeably in relation to undomesticated ducks that were in Australia before British colonial settlement in 1788.[56] While both native and imported game birds were protected under colonial laws, for native birds this was to "prevent destruction" by enforcing a closed season during the breeding season, and for imported birds so that they could establish and maintain populations for sport and food hunting.[57] These measures to protect overhunting did not always work, and there is strong evidence that in some regions of Victoria in the nineteenth century, hunters significantly reduced the populations of many native waterbirds.[58] In addition to open seasons, toward the end of the nineteenth century there was also some provision for protecting birds by conserving nesting areas, often linked to hunting interests. These "game reserves" were included in the Birds Protection Act 1881 and subsequent similar acts. The reserves could be declared for any of the listed birds, and in the twenty-year period following the introduction of this act, a series of wetlands were listed as reserves.[59]

Toward the end of the nineteenth century and into the early twentieth century, arguments for the protection of birds in Australia broadened to include their value to farming in addition to hunting interests, a shift that was reflected in the protection of insectivorous birds in various NSW state acts.[60] This change reflected the effects of an intense drought in the late 1880s and 1890s, which brought the economic and environmental effects of intense overstocking of sheep and widespread tree clearing by pastoralists into focus for both governments and farmers, and with it a realization that pastoral land use practices needed to change, partly because "native biota were suffering."[61] A now largely Australian-born Anglo-Celtic population had a better appreciation of native flora and fauna than previous generations of mostly European migrants, which may have contributed to this shift.[62] With these motivations, across this period some sectors of government and advocates of conservation looked to scientists for information on "the economic relationship that exists between animals, birds, and agriculture," partly to counter accusations by opponents of conservation that animal and bird protection was merely sentimental.[63] Arguments for the economic value of birds were arguments for their protection. Publication of scientific research into the economic role of birds in agriculture, for instance in eating invertebrates, proliferated in the 1920s and 1930s, and Kinghorn wrote about this topic beyond his work on ducks in rice fields.[64] People like Kinghorn saw "sportsmen" as allies in bird protection because they had "long realised the necessity of the protection of game during the breeding season."[65] As agriculture intensified and expanded in the early twentieth century, particularly with new irrigation networks and closer land settlement, including from soldier settlement schemes following World War I, problems between farmers and native birds and other fauna intensified.[66]

CONFLICTING INTERESTS AND THE EXPANSION OF RICE GROWING, 1930S TO 1950S

Rice growing expanded and became a fixture of farming in the MIA into the 1930s. Farmers' concerns over duck damage continued, and the NSW government's chief secretary declared "special open seasons" from 1 September to 31 December in rice areas each year from 1933 to 1938.[67] In 1938 ducks were noticeably "scarce" due to a prolonged dry period, and statewide open seasons were canceled in both Victoria and NSW.[68] Despite these general closed seasons, the NSW chief secretary declared a special open season on ducks in rice areas from 1 September.[69] Soon after the season began, several natural history societies led a protest to the government against the season having been declared at all because the months for the open season coincided with the duck breeding season. Representing the protesters to the *Sydney Morning Herald*, the secretary of the Royal Zoological Society (based in Sydney) argued that the chief secretary's decision was "high-handed and contrary to all humane ideas."[70]

This article generated a string of responses to the newspaper from farmers, zoologists, and ornithologists.[71] The debates remained focused on whether ducks should be hunted through their breeding season. In them, a division is evident between those in government agriculture, who argued for the elimination of ducks in the MIA (and whose views were reiterated by some farmers), and zoologists and naturalists, who were becoming increasingly aligned with species protection, and not just on economic grounds. It is perhaps in these debates that the views of some farmers about having ultimate control over what did and did not belong on their farms, and eliminating what did not fit, were articulated most directly. For instance, one farmer wrote, "I quite agree with agricultural instructor who said that 'all ducks should be destroyed,' meaning, of course, on the Murrumbidgee area."[72] In contrast, A. Basset Hull, the president of the Royal Zoological Society, repeated the society's "view that the opening of the duck season on 1 September was wrong" and also unnecessary, as Kinghorn's report had shown that "ducks are not a serious pest of rice crops."[73] This kind of interest from natural history societies in the conservation of native species was relatively new in Australia and only started in a sustained way in the twentieth century.[74] It is interesting to note too that in 1938 biologists made these arguments for the "humane" or ethical protection of ducks under the banner of the societies rather than in their government roles, which suggests that the societies provided an avenue for these arguments and interests that their government work did not.

The debates in the *Sydney Morning Herald* also provide a glimpse of the varied opinions and practices that existed among rice farmers. All of these centered on economics. For instance, one farmer devised a method for keeping ducks off rice during their nighttime feeding by lighting hurricane lamps, as he could not afford ammunition or shooting patrols, and he therefore argued that ducks could be deterred from rice bays instead of killing them. In contrast to such methods, other farmers argued that ducks needed to be eliminated in the irrigation area because of the damage they caused, estimated by one farmer as being £1,300 in a single season.[75]

A member of the Royal Ornithologists' Union raised the potential consequences of these open seasons on pastoralists, arguing that while the concerns of rice growers over damage from ducks "cannot be passed over lightly, as it represents big money," ducks needed to be protected for the benefit of wool growers, who represented "bigger money."[76] Ducks were helpful to pastoralists because they ate grasshoppers and snails that carried liver fluke, which could kill sheep.[77] The concerns of pastoralists and ornithologists highlight one of the many issues raised by the mobility of birds, as their treatment as agricultural pests in one area possibly undermined their ability to be agriculturally valuable in another. They also draw attention to escalating disputes between established pastoralists and newer irrigation industries as irrigated agriculture expanded.

The outcome of the protest by the natural history societies and the ensuing debates seems to have been that no special open season was declared in 1939, although the 1938 season continued. A farmer from the MIA lobbied biologists in 1940 to undertake further research into the issue, arguing that some farmers continued to be concerned that ducks damaged rice crops, while others argued that they helped the crop.[78] However, no research or open season followed. This may have been due to the view of at least some biologists, evident in the 1938 debates between farmers and natural history societies, that the matter had been resolved with Kinghorn's report.[79] It may also have been due to the outbreak of war.

Rice cultivation in the MIA expanded during WWII to supply international food aid to parts of Asia despite a severe drought in 1944 and 1945. Ducks were again attracted to rice fields during these dry years, and the federal government gave farmers some ammunition to shoot ducks. Ducks were also trapped by the mostly Italian prisoners of war who farmed the fields.[80] Following the end of WWII, Australia's international food aid program continued and, together with national food security, became a rationale for further increasing food production. Temporary wartime rice fields became fixed, and cultivation steadily expanded, in part sustained through soldier settlement.[81] The expansion of the area under irrigation was supported by new dams, particularly those built for the Snowy Mountains Hydro-Electric Scheme, which began construction in 1949.[82] In this postwar era, many local Wiradjuri returned to MIA farming areas as laborers for seasonal fruit picking and factory work, resettling parts of the towns established to service the irrigation industries, particularly Griffith.[83]

During the war years, the NSW government had continued to declare special open seasons, but on an occasional basis, for limited periods of time, and only for some areas.[84] This changed in 1952, when farmers attributed a rice shortage to duck damage. Following a visit to the rice areas, the chief secretary of NSW declared a special open season starting on 1 September 1952 and lasting for five months.[85] At the same time the Irrigation Research and Extension Committee, on behalf of the Ricegrowers' Association of

Australia, asked the Commonwealth Scientific and Industrial Research Organisation to investigate the "bionomics of wild ducks in the irrigation areas." In 1952 Harry Frith began this research.[86]

DUCKS AND THE EXPANSION OF RICE: FRITH

This was Frith's first assignment in the Wildlife Survey Section of the Commonwealth Scientific and Industrial Research Organisation (CSIRO), headed by Francis Ratcliffe since its creation in 1949.[87] Since 1946 Frith had been the assistant research officer at an Irrigation Research Station in Griffith, a town in the MIA. His work there had concentrated on the cultivation of orange trees, but he was increasingly fascinated by the birds of the region. In 1952 he sought to transfer to the Wildlife Survey Section to pursue this interest, as an opportunity to research ducks arose.[88]

Frith's research was requested by the Ricegrowers' Association, which had wholly disregarded Kinghorn's study because "his conclusions were based on two short visits to the Murrumbidgee Irrigation Area and an examination of very few ducks."[89] Frith's research was extensive, lasting a number of years: he drew on observations he had already made about the birds during his time in the region, as well as undertaking new research through 1956, which involved gizzard analysis of 1,849 ducks and field observations.[90] Like Kinghorn, Frith found "that a division of opinion existed among the growers themselves."[91]

Frith also found, like Kinghorn, that farming practices were important in limiting crop damage. He argued that the damage attributed to ducks from puddling the soils and "other activities" could often be traced to the farming practices described by Kinghorn, which led to patches where no rice germinated. In addition, he identified the problem of destruction from wind. Most significantly, Frith argued that bare patches in the crop from poor germination rates created good landing places for ducks, who could settle there and widen the opening by pushing down encircling plants. In a statement echoing Kinghorn's, Frith wrote, "It was observed that even well-grown crops of rice were very rarely visited by the birds." Both the MIA and the newer rice-growing regions of Murray valley irrigation districts were included in Frith's study, and he implied that farming practices, and thus damage, were worse in the latter because farmers were less experienced. He wrote that those in the MIA had been growing rice for "more than 30 years," and in contrast in the Murray valley, "rice growing was begun as a wartime expedient, and was undertaken largely by pastoralists with no experience of the crop"; though cultivation had "improved," Frith argued that "the area still lags far behind the MIA." For Frith, the consequences of farming practices on duck damage and the overall crop were his "most significant finding." Frith's argument was supported by the fact that special open seasons were only declared in the newer (and not the older) rice areas each year from 1953 to 1957 in response to complaints by farmers. Nevertheless, in both places Frith found that while different ducks fed on rice at different stages in the plants' growth, the damage was not extensive, and overall "wild ducks are not ... a serious pest of rice crops." An exception was the wood duck, which could graze down young rice plants if not scared off the bays, "but," Frith wrote, "such conditions existed usually only in a neglected crop."[92]

Another major aspect of Frith's work was linking duck population booms to floods. Floods occurred each year from 1950 to 1952, and again in 1955, allowing him to observe the birds' responses to these events.[93] The sheer number of ducks in the region during wet years, he suggested, led to more complaints rather than an increase in damage.[94] While there was a general understanding among biologists and farmers that ducks followed floods, Frith's research on their responses to floods laid much of the groundwork for later research by biologists on their nomadic behavior.[95]

Frith was aware that his research would influence whether further special open seasons would be declared. He argued that the belief "that unlimited and indiscriminate shooting in and around the irrigation areas would reduce the population of ducks or drive the birds elsewhere" was unfounded, as many species of ducks were highly mobile, and more ducks would appear from around eastern Australia. He also argued that hunters visiting the area during the special open seasons were more interested in "securing ducks than protecting rice fields." This led them to seek out ducks that were not on rice bays and then scare them onto the fields. Frith urged tighter controls on shooting and suggested hunters should be more like patrol groups.[96]

POSTWAR ECOLOGY AND CONSERVATION SCIENCE

Both Frith and his superior, Francis Ratcliffe, engaged with ecological ideas and helped foster the growth of conservation science in government institutions. Ratcliffe had studied zoology at Oxford University under Sir Julian Huxley, and from 1929 to 1930 worked as an economic entomologist in the CSIR (the precursor of the CSIRO), researching agricultural damage from flying foxes (or fruit bats) in Queensland.[97] In 1935, Ratcliffe began work with CSIR into widespread severe soil erosion that was devastating the pastoral industries in inland Australia. At the same time, the United States faced a similar problem with the Dust Bowl on the Great Plains. Environmental historian Libby Robin has argued that this crisis was a turning point in government science in both countries: "The soil erosion crisis in both the USA and Australia changed the emphasis of applied science . . . [as it] could not be handled on a 'pest control' model . . . but its progressive agenda increasingly emphasized development in the long term, not instant results." In Australia conservation science as it emerged in the 1930s–40s "became the next important umbrella for ecological work."[98] Although ecology has a long history as a science, with roots in the nineteenth century, it emerged as a professional discipline only in the postwar period, with strong ties to both zoology and the methods of field naturalists.[99] Ecology placed a central emphasis on the relationships between organisms and their broader community.[100]

In this context, the Wildlife Survey Section, led by Ratcliffe, was established to undertake a biological survey of Australia. However, the resources of the department were often claimed by the need for applied research, and Ratcliffe encouraged staff to incorporate ecology and conservation science into this research. Indeed, Ratcliff encouraged staff to combine ecology and agricultural economics during his management of the Wildlife Survey Section.[101]

Frith's research on rice fields sought to bring together the "ecology and economics of wild ducks," reflecting this philosophy.[102] Frith's ecological approach also meant that he followed the ducks off the farm, and he dedicated another paper to their breeding, food habits, and other aspects of behavior.[103] From his work on ducks, Frith argued that biologists needed to better understand the ducks in order to better know how to act in their long-term interest.[104] Later Frith wrote that "there are two main sorts of wildlife problems; those where the animal is a problem to man and those where man is a problem to the animal. The second is by far the more common and nearly always the more important."[105]

The intensification and expansion of agriculture in Australia after WWII raised concerns among biologists and more widely about reductions in habitat for native animals and birds. While new dams and irrigated farms were largely seen as central to national development, they also became a catalyst for an expansion in protected areas, which became a hallmark of conservation in this period, with its own sets of consequences.[106] This was reflected in the NSW Fauna Protection Act 1948, which placed a new emphasis on protecting habitat. The faunal reserves created under this act simultaneously created recreation areas for an increasingly urbanized population.[107] The act established a Fauna Protection Panel and brought together previously disparate laws about native birds and mammals, including pest control and wildlife protection.[108] In this and subsequent acts, including the 1967 and 1974 National Parks and Wildlife Acts, through which the NSW National Parks and Wildlife Service was created, birds and animals that colonists viewed as agricultural pests needed to be listed as unprotected, have open seasons declared, or have a permit issued before being killed.[109] Ducks were, and are, protected under state wildlife laws, except during declared open seasons. Starting in 1995 the NSW government placed a hold on statewide open seasons on ducks, effectively banning sport shooting in the state. This ban responded to increasing concerns among many groups about the ethics of hunting for sport.[110] However, from the time of Frith's report, special open seasons in rice areas continued to be declared by the NSW government in most years for several months across the Australian summer, as they still are.[111]

Because most farmers remain convinced that ducks damage crops, government ecologists have continued to pursue intermittent research into this issue. In 1997, a report from an investigation by Alison Curtin and Richard Kingsford indicated that there was "some belief" among rice famers that previous government research, including Kinghorn's and Frith's, had been undertaken when duck damage was not at its worst and so presented an inaccurate view of its potential severity. Further, in the 1970s biologists cast doubt on the accuracy of traditional gizzard analysis, throwing this aspect of Kinghorn's and Frith's research into doubt. Yet the 1997 report revealed a more complex picture than that presented by some farmers, suggesting, as other investigations had, that farming methods might be a factor in attracting ducks to rice fields.[112]

Australian conservation efforts have continued to emphasize habitat protection, in part due to international pressure. Beginning in the 1970s, Australia signed a range of international agreements for habitat conservation, including the Ramsar Convention in 1971.[113] This agreement aimed to stem the loss of wetlands, and consequently biodiversity, around the world, primarily for the conservation of waterbirds. Some wetlands in Australia have been included as Ramsar sites (including Fivebough and Tuckerbil

Swamps in the MIA), and their management has required the government to gain the cooperation of private landowners whose properties cover sections of or adjoin the wetlands, including some rice farmers.[114]

During the late twentieth century there was a shift in landscape ecology to understanding landscapes as mosaics.[115] This approach takes a holistic view that includes agricultural areas as habitats. Reflecting this change, in recent years some rice farmers have started to make more space for birds, animals, and plants, treating their farms as a kind of wetland. For example, in 2012, some organic rice farmers were also managing their farms "like a wetland," with birds and frogs helping with insect control.[116] Further, in recent decades Australasian bitterns, whose range is mainly Australia and New Zealand, have taken to nesting in rice farms in the MIA and elsewhere, finding these a more regular source of water and nesting material than other wetlands.[117] Concerned about declining numbers of these birds, from 2007 ecologists and conservation groups like BirdLife Australia have sought to work with rice farmers to raise awareness of the birds and encourage farmers to participate in their identification and conservation. For some of the ecologists involved, like Matthew Herring, this work also encourages farmers to rethink the boundaries and binaries between wildlife and agriculture by drawing their attention to the many animals their farms can help support. The Ricegrowers' Association of Australia has backed this project, and many farmers have provided photos and updates of Australasian bitterns on their farms, as well as giving the ecologists access to their farms.[118]

It is unclear whether farmers' interest in this sort of conservation is motivated by ecological/biodiversity goals or political agendas. Rice farming has come under increasing scrutiny in recent years for its traditionally high water use, the catalyst of which was the Millennium Drought (lasting from approximately 2000 to 2010 in eastern Australia). Starting in this period, the aquatic habitat created by the paddies has been used by the Ricegrowers' Association of Australia to help justify the crop, for example in publicly available brochures about biodiversity on rice farms.[119] There is at least an element of strategy in rice growers presenting their farms in this way during a period of heightened criticism. These efforts might be seen as a new negotiation by rice farmers, opening up the possibility of sharing these watery landscapes.

Ducks have not been considered in these negotiations. Special open seasons continue to be declared in the rice-growing areas. In their 1997 report Curtin and Kingsford suggested that duck decoy feeding areas could help draw ducks away from rice crops.[120] These are essentially dams with vegetation designed to appeal to ducks, much like the wildlife-friendly farm dams encouraged by various organizations and the NSW government over the last decade and a half.[121] While such ideas do not seem to have been widely taken up by farmers, perhaps because of water shortages during dry years, some farms now have wildlife-friendly dams. There are many conservation problems with these dams, such as their being on private land and subject to private interests. However, they also represent more hopeful possibilities for sharing these water landscapes—ones that perhaps both Kinghorn and Frith would have liked.

In this region, ducks have occupied a space at the interface of wildlife and agriculture. The multiple, sometimes conflicting, roles in which ducks have been cast reflect some of the complexity at this interface: as pests, game, economically beneficial animals, and protected native wildlife. In this way they participate in a larger and longer history of crossings that has also included flows of water and sewage, livestock and other wildlife, between farms and seemingly more wild places such as Fivebough and Tuckerbil Swamps. In the MIA, ducks in particular draw us into the diverse interests and understandings that have shaped crossings at the wildlife/agriculture interface. Kinghorn and Frith have provided focal points in tracing these controversies. Their work and professional contexts reveal some of the wider issues at stake in these controversies, including changing ideas about pests and conservation, and the shifting approaches in government biological sciences, from zoology and the economic conservation of species to ecology and conservation science. These relationships and interests have been, and continue to be, negotiated within a changed and changing water landscape.

Perhaps one of the most striking features of this case is that there has never been a consensus among farmers, or between farmers and biologists, about whether ducks significantly damage crops. These ongoing controversies highlight some of the diverse relationships among particular crops, farmers, and wildlife. Through these conflicts we can trace changing ideas about conservation and ecological relationships across various kinds of landscapes, including agricultural landscapes, protected swamps, and their wider hinterlands. Diverse imagined ecologies are taking form here, on and around rice farms, grounded in different understandings about these watery places and what and whom they might be good for.

FIVE

ENCLOSING

Pelicans, Protected Areas, and Private Property

IN MARCH 2015, I STOOD ON A ROCKY PENINSULA THAT JUTTED OUT into the Coorong lagoon in South Australia, sheltered from the wind by a bird hide. I squinted across the water as I tried to make out the features of an island in the lagoon that lay opposite. Australian pelicans were apparently breeding there now. In 1911, just over a century ago, this and other islands in the Coorong lagoon had been the site of a mass killing of young pelicans, likely by fishers who saw them as pests. The islands were subsequently leased by a group of ornithologists who sought to protect the pelicans and other birds that nested there by excluding fishers and local Ngarrindjeri Aboriginal people from the islands. Focusing on this event and its fallout, this chapter engages with the long history of contention over the enclosure of wetlands and its links with species protection, and considers the historical legacies of particular approaches. In this case we can see that these contentions and approaches were shaped by competing values and views among Ngarrindjeri people, fishers, ornithologists, and government administrators. Indeed, between 1908 and 1911 the legal status of pelicans on the islands changed back and forth several times, revealing competing values among individuals and groups regarding pelican protection and killing, even within the same government administration.

FIG 5.1 Australian pelicans near the northern end of the Coorong lagoon.
Photo: Emily O'Gorman, 2014.

Recent scholarship has sought to highlight the increasing involvement of nongovernmental organizations over the last few decades in the establishment and management of protected areas, as well as the increasing number of private protected areas in Australia and around the world. This trend has been examined as a neoliberalization of area protection, previously the domain of government agencies.[1] While the precise form this is currently taking—particularly in the rhetoric of sustainable development—may well represent a new paradigm of conservation, the pelican slaughter of 1911 and its fallout reveal the long history of involvement of nongovernmental groups not only in animal protection but in establishing private protected areas. In leasing the pelican breeding islands from the state government, the ornithologists attempted to enclose common lands, specifically to exclude both Aboriginal people and fishers. This can be understood within a broader history of the exclusion of Aboriginal people as well as other groups, often from lower classes, from private property and protected areas.

In the period following Australia's federation in 1901, governments increasingly established protected areas and national parks in response to the consolidation of a distinctive, racially based, white national identity that celebrated native fauna and flora, with contestations over who was included and excluded from this national imagery and these lands.[2] As others have noted, Australian Aboriginal people were not removed by governments and others from lands for the purpose of establishing protected areas, as Indigenous people in other countries were.[3] Yet this difference is marginal, as Aboriginal people were often removed from lands for other reasons and then excluded from protected areas.[4] However, the role of private protected areas in this history has been largely overlooked.

This chapter helps document the history and variety of area protection and exclusion, with particular attention to the role of private protected areas. Much more research needs to be undertaken into the history of private protected areas in broader dynamics of protected areas, and this chapter provides a situated study. These protected areas have helped create imagined ecologies of wetlands as enclosed and exclusionary spaces in Australia. Moreover, animals, and people's relationships with animals, were central to these processes. These histories are deeply more-than-human. In leasing the islands, the ornithologists aimed to reconfigure Aboriginal people's and fishers' relationships with the pelicans.

The regulatory dimensions of this history are crucial. Indeed this case demonstrates some of the complex ways in which places, as well as their human and nonhuman inhabitants, have shaped and been shaped by laws and regulatory structures. While a range of individuals and species were involved, my discussion here primarily centers on two groups: the pelicans that nested on the islands and the ornithologists who sought to protect them. I first came across this event through an archive in the State Records of South Australia that documented the leasing of the islands to the ornithologists. This archive reveals the complex roles of both the law and the ornithologists, as the ornithologists sought to maneuver through a legalized landscape to find ways around legislation that did not offer the kind of protection they wanted for birds, with mixed outcomes for all involved. In this chapter I draw on this archive as well as other sources, such as newspapers and ornithological publications, critically placing them within broader colonial power structures and the discourses in which they were generated, and using them to both examine the ornithologists' complex position over pelican protection and provide a lens onto wider issues and other values, namely those of fishers and Ngarrindjeri people.[5]

The slaughter reveals longer histories of conflicting ideas about killing, protection, and private property in Australia and internationally, which shaped the lives of both humans and nonhumans. Specifically, it illuminates the history of the protection and persecution of particular species at a time when bird and animal protection was gaining increasing support nationally and internationally. Further, the long history of class conflict over the exclusion of particular groups from private property intersected with and shaped bird protection laws, as well as conflicting views of hunting laws and bird sanctuaries. The second half of the chapter specifically focuses on why government officials and others saw leasing the islands to the ornithologists as offering greater protection for the pelicans than the available legislation, as well as how this became tied to ornithologists' discussions over limiting Ngarrindjeri people's collection of waterfowl eggs. Ultimately, this chapter contributes to historical perspectives on how people and animals have lived in landscapes with conflicting values, what has been at stake in these contestations, and the uneven consequences of their outcomes.

Pelicans, islands, and lagoons help us understand the law. In many ways they all helped shape legal aspects of the events under consideration here—for instance, the pelicans' selection of the islands as a rookery site and their colonial nesting patterns, which were linked to the water and fish of the Coorong. Indeed, the very size and appearance of pelicans were central factors in this controversy, as this meant for some that they were conspicuous competitors for fish and for others a loved icon of the region and Australia that was in need of protection.

THE COORONG, PELICANS, AND ISLANDS

Today, the Coorong is an iconic wetland in Australia, known for its birdlife. The area has historically attracted many kinds of birds, with over two hundred species recorded in the region, in addition to other animals and plants.[6] In 1966 part of the region was declared a national park, and in 1968 a separate area encompassing many of the islands was declared a game reserve, with both areas coming under joint management of the National Parks and Wildlife Service in 1972. In 1985 the Coorong together with Lake Albert, Lake Alexandrina, and the mouth of the Murray River were listed as a wetland of international importance, particularly as a waterfowl habitat, under

FIG 5.2 The Australian pelican. Henry Constantine Richter and John Gould, *Pelecanus conspicillatus Temm.*, Australian pelican, 1847, http://nla.gov.au/nla.obj-139482723.

the Ramsar Convention. In the 1980s, the Coorong further became subject to migratory bird treaties with Japan and China.[7] In the twentieth century, the Australian pelican has been just one of the birds that regularly visit the Coorong, albeit a visually prominent one, as these birds can number in the thousands.[8] They have a distinctive look, with large pink bills and black-and-white plumage, and they are big. In fact, Australian pelicans have been classed as one of the "heaviest flying birds in the world," with adult males on average weighing eight kilograms.[9] It is perhaps no surprise, then, that the specific epithet in the species name, *Pelecanus conspicillatus*, means "conspicuous." According to biologist Julian Reid, nesting adult pelicans need to eat approximately 0.4–0.5 kilograms of fish a day, and the availability of this food source has been a major factor in the location and size of rookeries.[10] For this reason, pelicans often use islands as rookery sites, offering some protection from predators for eggs and young birds.

Nori (pelicans) are important in Ngarrindjeri songlines; they are moral actors and ancestors. According to one account, pelicans and other fishing birds, who were once men, shaped the characteristic claypan and samphire swamp landscape of the wider Coorong area as they padded down the shrubby vegetation (mostly mallee bushland with their feet. Magpie, a land bird, refused to share fire with the other birds, so all they gave magpie from their fishing nets was silver bream, a bony fish. This caused a fight in which crow's eyes were burned black by the fire and pelican's previously black body was splashed with white scales from a silver bream. In another account, the fight is between land and water birds, with magpie leading the former and pelican the latter. This fight still constantly threatens to erupt. Eventually, the pelican-man dove into the water, taking drum nets with him, and transformed into a pelican, with the nets forming the pouch-like beak.[11]

Like many Australian waterbirds, pelicans are nomadic rather than migratory and have been able to respond to sometimes drastic changes in water and fish availability. Many thousands of birds (numbering over one hundred thousand) have been known to suddenly congregate on floodwater in the normally arid interior of the continent. These floods can fill large temporary lakes, such as Lake Eyre in central Australia, and pelicans have exploited the safe rookery sites offered by islands that form in the floodwater to court and nest, feeding on the increased fish stocks of bony herring (or bony bream). Reid has noted that how pelicans know when and where to migrate during these times of flood remains "one of the great unsolved natural history mysteries." The prolonged breeding that takes place during these floods has occasionally radically changed both the numbers and the distribution of pelicans throughout the continent. For example, during large floods in 1974–76, thousands of pelicans gathered at Lake Eyre, and as the birds dispersed with the floodwaters, pelicans became a common sight where they had previously been rare.[12]

For pelicans, one of the reasons the Coorong has been, and continues to be, significant is because of its islands. Pelicans have nested on the "Pelican Islands"—a group of six limestone islands—almost every year since records began to be kept in the colonial period, at times laying as many as 3,500 eggs.[13] "Pelican Islands" is an informal name colonists gave to a group of six limestone islands.[14] Pelicans have been known to breed there any time between June and March, and biologists have linked the variation in breeding times to changes in water height.[15] By the early twentieth century ornithologists regarded the islands as one of two main nesting places for pelicans in the state.[16] Another once existed on an island in Pelican Lagoon, located on nearby Kangaroo Island. However, this ceased to be used in the late nineteenth century, possibly due to raids or hunting pressure.[17]

Aboriginal laws stem from relationships within particular Country (for example, among particular people, plants, animals, land, and water), and areas have primarily been the territory of certain groups.[18] The Coorong has been territory of the Tanganekald (or Tanganalun), a group within the Ngarrindjeri. Early in British colonization, which officially began in South Australia in the 1830s, Ngarrindjeri numbers significantly decreased due mostly to introduced diseases such as smallpox and frontier conflict with colonists.[19] According to one source, in 1842 there were about 3,200 Ngarrindjeri, and in 1874 only 511.[20] The British government had declared South Australia to be "unoccupied" prior to official colonization, and in the late nineteenth and early twentieth centuries many Ngarrindjeri were moved onto land set aside by the government for Aboriginal people and onto pastoral properties and missions.[21]

A small fishing industry started in the Coorong and Lower Lakes region in the 1840s, expanding rapidly after 1885, when steam rail connected the local town of Goolwa to Adelaide markets.[22] Commercial fishing shaped local settlements economically and socially, providing employment and supporting the development of related industries. In the period 1908–13, fishers from the Coorong were supplying 11–14.3 percent of the state's annual fish haul of 1,504.6 tons, mostly yellow-eye mullet, tommy ruff (Australian herring), and mulloway.[23] Local fishers have long seen pelicans as pests, and reports of young pelicans being killed and eggs smashed on the islands in the Coorong go back to at least the 1870s.[24] It is unclear how often these raids on the rookeries occurred; some evidence suggests they may have happened almost every year. In 1963, an ornithological paper on pelicans in the Coorong region noted that these massacres had apparently happened often enough that "the occasional mention of successful breeding in the early records may imply that raids did not occur in that season."[25] In many ways, then, the 1911 slaughter was not an unusual event. However, the strength of the debates it mobilized, particularly regarding bird protection, the archival and newspaper records it generated, and its fallout make clear that something set it apart. In many ways this something may be the particular historical moment—locally and more widely—in which the slaughter took place. Further, the strength of the debates and the fallout seem to indicate that more birds were killed in the 1911 massacre than in previous years. This was likely because pelicans were no longer protected, and a bounty had been put on their heads, introducing a financial benefit to killing them. I have found no evidence of a previous bounty on pelicans in this region, though there had been rumors of one. Chapman noted that in 1903 "many people in the Lake Albert and Lake Alexandrina area were under the impression that the Government was paying scalp money for pelicans," but this was denied by the government.[26] The 1911 slaughter, then, occurred in a local context of past raids and wider government efforts to limit damage from pests to both fisheries and agriculture, as well as a growing protection ethos among various individuals and groups.

THE PELICAN SLAUGHTER

In February 1911, news began to reach Adelaide of the slaughter on the Pelican Islands. A local resident of the Coorong area, F. G. Ayres, wrote to the *Daily Herald*, a metropolitan newspaper published in Adelaide, to alert "the bird-loving public of South Australia to the law which permits the destruction of pelicans." In his letter he recounted the killings: "Recently a small party of men from [the nearby town of] Meningie raided the pelican rookeries on the Coorong and as a result returned with over 2000 pelican heads, for which they received payment from the Meningie police on behalf of the Government." He blamed the government and not the men for "such scandalous destruction of one of the most majestic birds in Australia," as it was the government that had put the bounty on the pelicans. He was concerned that the putrefying pelican carcasses would have deadly consequences for other birds nesting on the islands, including terns, gannets, and silver gulls, as adults could possibly abandon their eggs and young, which were now surrounded by rotting flesh. He also argued that pelicans were being unnecessarily persecuted and that other fish-eating birds were nonetheless protected: "Are not the gulls, the gannets, the various terns, herons, bitterns, griebes [sic], and the beautiful egrets voracious feeders on fish also? Yet the destruction of one of those birds entails a penalty of £5."[27]

The law that Ayres was trying to call attention to was the Fisheries Act Amendment 1909, which had been passed by the South Australian government in November 1909 and came into force in early 1910. This legislation, which amended the Fisheries Act 1904, legally declared pelicans a pest to fisheries in the state. The amendment allowed inspectors, if authorized by the state fisheries minister, to "receive and destroy the heads of . . . pelicans."[28] Other animals that had been included as pests in the original act of

1904 stayed on this list, including cormorants, turtles, and tortoises.[29] Many in the fishing industry saw these as "fish enemies," or more precisely, as enemies of fishers. They were blamed for eating, and thereby reducing hauls of, marketable fish, and consequently the minister could declare a bounty on them.[30] Pelicans were particularly conspicuous competitors for fish, as they took fish from the nets, often when fishers brought the catch to the surface.

However, at the same time as pelicans were officially listed as pests, they were also protected for part of the year (July to December) in the state under the Bird Protection Act 1900.[31] This act had three schedules: the small group of birds listed in the first schedule were protected throughout the state all year; the second schedule included the majority of birds (including pelicans) and specified closed seasons, intended to protect birds during their breeding season from both market and sport hunters, both of which were frequent visitors to the Coorong; and the third schedule listed birds that were not protected at all.[32] Until recently, pelicans on the Pelican Islands in the Coorong had been protected year round through another legal mechanism. In 1908 the governor had declared part of the region, which encompassed the islands, a "bird protection district" under the same act for "a period of five years."[33] This protected birds in the area listed in the first and second schedules of the act for the whole year, but not those in the third.[34]

The protected status of pelicans on the islands, and throughout the state, changed again in 1910, when the state government removed pelicans from the second schedule and placed them on the third.[35] This meant that there was now no closed season on killing them, and they were no longer protected in the protection district. Presumably, pelicans were moved to this schedule in accordance with their listing as a pest to fisheries in 1909. Further, the minister declared a "bonus of a penny per head" on pelicans, which was paid from the fees for fishing licenses.[36] While there were a number of legal factors and interests at play, many regarded the bounty as the direct motivation for the slaughter. This was supported by a tally from the chief inspector of fisheries that "bonus' [sic] have been paid on 2500 heads."[37] Those who undertook the slaughter did it in such a way as to claim the bounty: they waited for birds to hatch in order to collect their heads, as they needed to present these to authorized officials to collect the bounty.[38] At a penny each, the total payment for two thousand pelican heads was £8 3s, equivalent to A$1,333.20 in 2019.[39] This was a reasonably large sum of money for what would have been a few days' work at most. Given the view of many fishers that pelicans were a pest and the fact that there had been past raids, those who did the killing may have also seen this as a community service to the local fishing industry. The bounty encouraged people to kill pelicans, reinforcing it as a "good" or community-minded action. However, as news of the slaughter spread, many people were outraged that young birds had been massacred and questioned whether pelicans constituted a pest at all.

That Ayres wrote his letter to a metropolitan newspaper is significant. It implies that he was trying to bring these circumstances to the attention of those involved in bird protection in Adelaide. Prominent among these were the members of the South Australian Ornithological Association (SOA). The SOA was a strong advocate of bird protection and had been involved in lobbying for a range of bird protection districts to be established in the state.[40] Two leading figures in the SOA, Samuel White and John Mellor, were particularly active in responding to the slaughter; earlier Mellor had publicly voiced his opposition to the removal of pelicans from the list of protected birds.[41]

White expressed his view of the slaughter to newspapers soon after the publication of Ayres's letter, stating, "Some scoundrels, it appears, allowed the whole rookery on Pelican Island [sic] to hatch, to the extent of 2000 birds. The rookery has now been swept away because somebody has secured the heads of 2000 fledglings.... It is one of the most dastardly acts I have ever heard of." White argued that more such raids—"so brutally perpetrated"—would lead to "the extermination of this remarkable bird," as this was "one of only two rookeries in South Australia." Further, like Ayres, White viewed the bounty and removal of protection as a vilification of pelicans that would not achieve the government's desired result of increasing fish hauls. He argued that "we [the SOA] can prove that pelicans do not consume the enormous quantities of fish they are alleged to do."[42] Indeed, other biologists and members of the SOA similarly argued that pelicans did not eat enough fish to be pests, or that they ate fish like bony herring, which commercial fishers did not want.[43] White asserted that it was not just pelicans who were targeted unfairly but also cormorants, and that instead of increasing fish hauls, "where cormorants have been destroyed fish have diminished in number."[44] There was a long history of persecution of fish-eating birds, particularly cormorants, in the United States and Australia, and longer still in England.[45] White argued that this traditional view of fish-eating birds needed rethinking. He drew parallels with the United States, where pelican numbers had plummeted after similarly being targeted as competitors for fish. However, White stated, "about 3 or 4 years ago the authorities awoke to the fact that these birds were not injurious to the fishermen's trade, and the authorities set to work to reestablish the birds" by proclaiming an island where pelicans could still be found as "a bird sanctuary."[46]

Ornithologists and others saw the pelican massacre as possibly contributing to growing instances of local and species extinctions of birds in Australia and around the world. In the North Atlantic, great auks had become extinct in 1844, and emus and pelicans as well as other birds could no longer be found in some areas of Australia.[47] Following the slaughter on the Pelican Islands, one reporter wrote that "the massacre had amounted to thousands, and this would soon exterminate the quaint bird, of which Australia had only the one species."[48] Indeed, many reporters commented on Australian pelicans as "quaint," "noble," and "remarkable" birds that were native to the continent, and so in need of protection.[49] For instance, one reporter wrote that the pelican was that "peculiar pouched bird which in Australia is the handsomest of its species."[50] This resonated with the new nationalism that took root in Australia following federation and celebrated native fauna and flora.

The story of the raid on the rookeries, and particularly White's interview, was picked up by metropolitan newspapers across eastern Australia and in some regional presses. The massacre of young birds seemed insupportable to many, particularly when coupled with White's analysis that pelicans in fact were not eating the quantities of fish that the fishing industry claimed. The reporter for the *Daily Herald*, for example, which published White's views at length, wrote that "the most enthusiastic fishermen . . . will object strongly to the tactics commented on in the letter [by Ayres], and will welcome the day when such cruelty is no longer possible."[51] This reporter also called the massacre of pelicans "illogical," and another deemed it both "foolish" and "cruel."[52] Many labeled the bounty "blood money."[53]

While no one defended killing young birds, local fishers voiced their views that pelicans reduced their hauls, eating the fish they wanted for market. A former fisherman from the town of Meningie, W. Tregilges, wrote to the *Daily Herald* that "I have frequently been four or five dozen . . . bream short [due to pelicans]. . . . I have put out a mullet net at night and in the morning have seen about 20 or more of them [pelicans] along the net quietly saving me the trouble of taking the fish out, but they would go a little further than that and cause me to buy more nets" because of the damage they caused when they pulled the fish out.[54] For Tregilges, while "the pelican, as Mr. Ayres says, may be a majestic bird from his and a few other's [*sic*] point of view... according to my idea it is one of the most useless and ugly birds we have."[55] Fishers, ornithologists, and others disputed knowledge about pelican behavior, specifically whether they ate large quantities of marketable fish. These conflicting knowledges may reflect the differing values between these groups, which could fundamentally frame how they defined the problem.[56] For example, these groups may have held different views about what were acceptable losses: what ornithologists regarded as minor losses, fishers may have seen as major or unacceptable, with the added expense or inconvenience of damaged nets.[57] There were also clear socioeconomic differences intersecting with and shaping these different values and needs, between the leaders of the SOA, who advocated for pelican protection and were from middle-class or wealthy backgrounds, and the fishers, who were often poorer. This division also cut another way, between the city-based SOA leadership and rural fishers. These differences in background likely influenced the contestation between the two groups in a broader sense (although never explicitly). To fully understand this event and its fallout, it is important to place it in a longer and wider context of contested values and ideas about protection, killing, and private property.

PROTECTION, PESTS, AND PRIVATE PROPERTY

The pelican slaughter and its fallout were shaped by contested ideas and values at a range of scales, including international. Environmental historian Thomas Dunlap noted of animal bounties in the Anglo-European settlements of the United States, Canada, and Australia, "Everywhere the settlers had the same goals and used the same technologies, but the lands and the animals were different."[58] These differences and similarities were important in shaping local responses. Many other laws concerning animals and contestations over the role of private property also resonated in these places, and at times in places such as India and Africa, which shared various cultural and economic links, including through the British Empire.[59]

In Australia, early legislation for the protection of animals was part of the wider project of acclimatization, and indeed colonization, undertaken by governments and private citizens throughout the British Empire and the United States.[60] These laws often sought to establish useful or valuable European game species, such as pheasants or rabbits, while protecting Australian native game from overhunting—for example, water birds like native ducks and black swans, which took on the role of similar game species in Europe.[61]

In the late nineteenth and early twentieth centuries, the focus of Australian protective legislation changed, as most states passed legislation that offered some protection to native animals by default, and unprotected animals needed to be listed. South Australia did this in 1874, which is relatively early in Australia.[62] For example, NSW only passed a similarly inclusive act for birds in 1918.[63] This shift reflected people's concern that many kinds of native birds were being hunted for various reasons (not just "gentlemanly sport") and were in danger of being exterminated altogether.[64] Colonists killed native birds, for example, as pests, for the millinery trade, and for market, and others were often killed accidently during these activities. In addition, birds were routinely captured to trade as exotic pets to overseas customers or killed as trophies.[65] Pelicans, for instance, were killed by market hunters and at times sport hunters for their skins and plumage.[66] The protection laws mainly aimed to regulate hunting through closed seasons over the period that was thought to be the birds' breeding seasons. Yet the laws that aimed to protect birds and other animals were often ignored, and many people killed them year round, as with the pelican rookery raids on the islands in the Coorong.[67]

While these acts marked a shift in bird protection, some viewed them as a continuation of British game laws, and in South Australia politicians debated them at length. In Britain, game acts were passed starting in the late eighteenth century that entrenched divisions and heightened tensions between socioeconomic classes by giving "the landholder the right to protect them [the animals] against poaching or trespass."[68] Effectively, this meant that those who owned land could kill game on their land year round, and those who were not landowners could not. Those who caught or killed game on other people's land could be heavily fined and, if caught at night, imprisoned. In Australia, many people wanted to prevent similar game laws from being implemented. Yet in South Australia these protection laws were seen by some to similarly favor landowners because they could do as they liked on their own land while others were excluded, and killing animals on crown land was heavily regulated. Aboriginal people were exempt from these laws if they were hunting or collecting eggs on crown land (or "waste land" in earlier legislation) but not on private land unless they first gained the permission of the owner.[69]

Animals and birds are important in Aboriginal kinship and practices. In the Coorong region, pelicans were important to Ngarrindjeri people as one of the main *Ngaitji* (creation ancestors). Their eggs were collected from the islands for food, the birds were eaten, the fat was used to strengthen plant fibers, and the feathers for hunting equipment or, in this period, feather flowers.[70] While the exemption of Aboriginal people from the acts (which meant they could hunt year round) might be seen as a recognition of the importance of hunting and egg collecting in sustaining Aboriginal communities, it is perhaps better understood in relation to contested colonial legal views, which, as noted by Ngarrindjeri scholar Irene Watson, centered on whether Aboriginal people could be tried under colonial laws.[71] In the colonial period in South Australia, some colonists viewed Aboriginal people as "wild and savage," with very different laws and unable to be adequately tried under British laws.[72] Others thought those who were "making advance towards civilization" should be tried as "British subjects."[73] Aboriginal people's exemption from bird protection laws reflects these contentions about their inclusion in this legal system. The bird protection laws also reinforced the ongoing dispossession of Aboriginal people since colonization by excluding them from privatized land.

Alongside these protective acts were those that aimed to destroy animals and birds seen as pests. In Australia, by the 1880s there was an extensive system of bounties for particular native and introduced animals and birds, supported by legislation.[74] Many of these, including those in South Australia, included provision for "scalp money" on birds and animals that were declared pests to crop and livestock farmers.[75] In South Australia, it was only with the Fisheries Act 1904 that animals regarded as pests by fishers were officially listed as such.

This bounty system was part of what many, as historian Penny Olsen wrote, "have ... described as the fauna wars of the late nineteenth and early twentieth centuries," in which many animals and birds were killed because they were seen as competing for resources or were blamed for stock, crop, and fish losses.[76] The killing undertaken in this period was extensive and often fixated on eliminating particular species in a given area. Deborah Bird Rose, in her work on past and more recent dingo baiting, has used the phrase "will-to-destruction" to describe this systematic killing of all of a particular kind of animal.[77] A single pest animal or bird was seen as too many by some farmers and fishers (but not all) and any losses of livestock, crops, or marketable fish as unacceptable.

In this context, from the end of the nineteenth century into the early twentieth century in eastern Australia, arguments for the protection of birds as valuable to farming and fishing, in addition to hunting interests, gained increasing traction among biologists and others.[78] As noted previously, in NSW these changing views were underpinned by the effects on animals of a severe drought at the end of the nineteenth century, exacerbated by intensive farming practices and a growing appreciation for native fauna by the now largely native-born Anglo-Celtic population. Taking a national perspective, Steven White has argued that this new emphasis on protecting native fauna was partly due to a new sense of national identity that emerged in the 1880s and 1890s as discussions about federation gathered pace, along with the emergence of "specialised animal advocacy groups" in response to the "destruction being wrought," the failure of many introduced species to adapt to Australian conditions, and a larger Australian-born population.[79] Internationally, movements to protect birds were growing. In Britain, the United States, New Zealand, and Canada, advocacy groups were forming to protest the killing of birds for a range of reasons, including the use of bird skins and feathers in women's fashion, a trend that resulted in the death of millions of birds from around the world each year.[80] These groups created international networks, for example the British Society for the Protection of Birds, which protested the plume trade and had branches in many countries, including Australia.[81] With these motivations, in many places during this period government departments and advocates argued for the protection of birds based on their utility in agriculture and fisheries (because they ate pests), partly to avoid being dismissed as sentimental.[82]

As R. Crompton noted in his history of ornithology published in 1914, the main goal of the South Australian Ornithological Association (established in 1899) was to protect "useful" birds: "This Association has since its inception stood:—Firstly, for the protection of our native birds ... as many of them are insectivorous, making their principal if not their only food of pests.... Others ... consume enormous quantities of the seeds of thistles and other weeds.... Others again are scavengers, eating dead animals, thereby arresting the spread of disease among our stock. Again others are the enemies of snakes, snails, etc., and lastly the cormorants, feed on the enemies of our fish although fishermen do not all recognise it."[83] The SOA championed the Bird Protection Act 1900, which reflected many of the members' interests, including a focus on protecting economically useful birds and regulating the live export trade and plume hunters. According to Crompton, members of the SOA had to "fight the hard fight for the inclusion of almost every species," particularly those seen as pests, because the members of parliament "could only be apprised of the good they do, by a body of ornithologists."[84] The bird protection districts established by the act were "considered by a certain section of the House to be a copy of the British Game Laws which they hoped would never be introduced into Australia," but the act nevertheless passed.[85] Support for humanitarian protection of birds and animals grew through the first decades of the twentieth century and was an explicit reason for animal protection in the South Australian Animals Protection Act 1912.[86] Indeed, these sentiments were evident in the widespread condemnation of the slaughter of young pelicans in 1911.

PRIVATIZING PROTECTION

Following the slaughter, in May 1911, the SOA organized a deputation—including Mellor and White as well as a number of other advocates of bird and animal protection—to the commissioner of crown lands, Crawford Vaughan, to request that pelicans again be given partial protection by being returned to the second schedule of birds. Vaughan, however, had other ideas. Instead, he had been "thinking of establishing . . . a defined area on the Coorong of absolute protection for birds in the breeding season," partly to protect them from sport hunters who "'accidentally' hit the protected birds" on shooting trips.[87] By the turn of the century, the Coorong had become well known for its birdlife and as a destination for groups of recreational shooters from Adelaide.[88] Vaughan saw it as a duty of the government to enforce the protection of the state's flora and fauna, "as Australia had the most peculiar fauna and flora in the world," and he promised to pursue the possibility of establishing a sanctuary for all native birds.[89]

Bird protection legislation, however, did not absolutely protect all native birds, as birds listed in the third schedule were unprotected even in sanctuaries. Nor could it limit people's access to crown land. The solution Vaughan used was to lease a number of the islands, including the Pelican Islands, to the SOA on the condition that they prevent people from visiting them.[90] This idea resembles other creative solutions used by bird protectionists in this period, which aimed to sidestep legal obstacles. One example on an international scale is the Migratory Bird Treaty of 1916 between Canada and the United States, which sought to overcome both state opposition to federal bird protection laws in the United States and concerns that such laws were unconstitutional, by bringing bird protection into the clearly federal realm of treaties.[91] With the lease, which took effect in August 1911, the Crown

Lands Office gave the SOA the right to "prevent any person or persons from trespassing," a key condition of private property, and the SOA had soon erected signs notifying people to keep off the islands.[92] Over the next few years the SOA's management of the islands was aided by a local resident, William Goldfinch, "who knows all the nooks & crannies of the Coorong" and who was paid by the Crown Lands Office to patrol the islands for trespassers by boat and to kill crows as well as foxes and rabbits—no longer seen as game to introduce and protect, as they had been in the colonial era—to protect the pelicans.[93] By way of comparison, protecting pelicans in this way led to a more complex outcome on the Coorong than on the Swan River in Western Australia. There, similar disputes at around the same time over fish-eating birds were ultimately won by developers and the fishing lobby, which overrode ornithologists' efforts to change legislation to protect the birds. The ornithologists there had no last-minute creative legal maneuver, as on the Coorong.[94] It seems that ornithologists' political strength in South Australia, as well as the presence of an empathetic and powerful government official in Vaughn, influenced the different outcome on the Coorong.

Aboriginal peoples' collection of the eggs of both black swans and pelicans became entangled in these responses to the pelican slaughter on the Coorong, particularly from the SOA. Ornithologists and naturalists did not agree with Aboriginal people's collection of waterfowl eggs, which stemmed from pre-European practices.[95] One intention of leasing the islands appears to have been to stop Aboriginal people from gathering these eggs. Charles Barrett, a naturalist and coeditor of *Emu* (the journal of the national Royal Australian Ornithological Union), wrote a short statement in one of the volumes, drawing attention to this motive: "S. A. White . . . has been making strenuous efforts for the last twelve months to secure legislation for the protection of the Pelicans and to prevent the so-called aborigines from robbing the nests of Black Swans and Pelicans. The name of the R.A.O.U. [Royal Australian Ornithologists' Union] has been used in urging such legislation. . . . The island rookeries will now [that they have been leased], thanks to the efforts of the South Australian Ornithological Association be less liable to receive visits from the bird-killers and egg-robbers."[96] In the context of concerns over the effects of bounties and raids on these rookeries by fishers, as well as longer histories of killing fauna since colonization, Aboriginal peoples' collection of eggs was seen by ornithologists and others as damaging to pelicans and swans.[97] While the SOA and others viewed leasing the islands as an immediate measure to prevent

Aboriginal people from gathering pelican and swan eggs on these islands by making these come under some of the laws of private property, changing the legislation to prevent these practices more widely seems to have been a broader goal of at least some ornithologists, such as White and Mellor. Yet Ngarrindjeri people have protocols about the collection of eggs and the treatment of pelicans and swans. For instance, some Ngarrindjeri have said that they leave two swan eggs in each nest, and they distinguish between fresh eggs and those about to hatch.[98] The ornithologists seem to have been unaware of such protocols at the time.

In the month following the leasing of the islands, Mellor traveled to the Coorong on behalf of the SOA, along with the local inspector of fisheries, M. C. Ewens. In his account of the visit to the commissioner, he wrote that the "natives rob the nests disgracefully, taking both fresh & well incubated eggs.... [T]he latter are thrown out.... I have the records from authentic sources that the natives go in small parties ... to the best breeding places of Lakes Alexandrina & Albert, & the Coorong, take hauls of 200, 400, & 500 eggs of the swans, this is repeated as long as the laying lasts.... [T]he Bird Protection Act [should] be applied to blacks and whites from the line south of Adelaide and Mannum [a town at approximately the same latitude]."[99] Numerous scholars have highlighted that a continent-wide Aboriginal identity in Australia only formed in the context of colonization and "black" and "white" relations.[100] The category of and identification as an Aboriginal person emerged in a shifting social and political terrain, and remains a contested notion in Indigenous communities.[101] Mellor seems to have had a particular, narrow understanding of Aboriginality, which he thought no longer applied to some in the Coorong region. In arguing that Aboriginal people should be subject to the bird protection laws, Mellor wrote that "both full blooded blacks & half castes" took eggs, and that "the blacks have rations given to them, & these parts are now fairly well civilized."[102] Indeed, Barrett's comment that "so-called aborigines" took eggs may have stemmed from similar views. Mellor further suggested that rations given by the government to local Aboriginal people could be "increased as compensation."[103] These statements echoed the complementary discourses of colonization, race, and assimilation that gained increasing traction in Australia in the nineteenth century and endured into the twentieth century.[104]

In his letter to the *Daily Herald*, Tregilges, the local fisher, also raised the issue of the exemption of Aboriginal people from bird protection laws. He questioned the protection of swans, which were listed in the second schedule, as they were numerous in the Coorong area: "I cannot see why the white man should be debarred from taking a few of their eggs to eat and the natives allowed a free hand." Tregilges argued that Aboriginal people had collected about two thousand swan eggs, many of which were about to hatch: "Now a white man would not do a thing like that." Here Tregilges, like Mellor, drew on hierarchical ideologies of race. Tregilges included the treatment of birds in this, racially vilifying Aboriginal people by portraying them as cruel to birds and thus "uncivilized."[105]

Echoing these discourses, the state Animals Protection Act 1912 stated that only "full-blooded" Aboriginal people were exempt from bird protection legislation. While the Animals and Birds Protection Act 1919 did not include this qualification, it did include the paternalistic provision that if "any of the privileges ... are being abused," the governor could suspend them.[106] The National Parks and Wildlife Act 1972 did not include any provisions for Aboriginal hunting or egg collecting, and Philip Clarke noted in his cultural geography of the region that after this act the "swan-egging practices of the local Aboriginal people were by stealth."[107] Many state acts of the 1960s and 1970s did not include exceptions for Aboriginal people and therefore prevented activities such as hunting, burning, and harvesting plant material in protected areas. While this was also the case for Coorong National Park from 1972, Aboriginal rangers have facilitated the incorporation of some of these activities into park management in recent decades.[108] Over the last forty years various avenues have been developed at state and national levels through which to include Indigenous people in protected areas management, a process that has been significantly influenced by the Aboriginal land rights movement.[109] Planning is currently underway for Ngarrindjeri comanagement of Coorong National Park.

PELICANS IN A CHANGING LANDSCAPE

In his 1911 letter to the crown lands commissioner, Mellor reported on how the pelicans were faring. He wrote that "the poor old pelican is still breeding on several of the islands under the control of the Assoc. This is gratifying, as I believe that this is the only spot in this state where this particular bird now breeds."[110] A few months after the slaughter, pelicans had returned to at least some of the islands. There had been past slaughters, and current understanding of pelican biology tells us that pelican chicks have high mortality rates even without people killing large numbers of them, so perhaps the slaughter of 1911 was not as significant to the pelicans as it was to the ornithologists. SOA's leasing of the islands, however, did not stop raids from happening altogether, and people continued to slaughter the young birds and smash eggs. For example, in 1941 ornithologist H. T. Condon wrote that "many times during the last ten or twelve years the pelicans on the islands of the Coorong, South Australia, have been subjected to marauding parties which have clubbed the young ones to death or trampled them down with heavy sea-boots." He noted that it was difficult to prosecute those who undertook the raids because "informants are always unwilling to state their evidence publicly."[111] In 1963, Chapman wrote that across the previous sixty years there had been a "marked decline" in the number of pelicans in the region, but it was difficult to quantify because recordings of the number of pelicans had been sporadic and often made during infrequent visits by different biologists.[112] He attributed the general decline partly to the raids on the rookeries but also to various changes in the lagoon and the region more generally over this short period. These included the draining of swamplands for agriculture both locally and along the length of the Murray River, the canalization of this river for shipping and irrigation, the construction of barrages blocking the mouth of the Murray River, and the introduction and spread of new species of fish in the Murray system. Some of these changes increased the salinity and altered the depth of the lagoon, as well as changing other aspects of pelicans' habitat and food resources. After a widely publicized raid on the rookeries in 1962, the state government regained management of the islands, declaring them "prohibited areas" and placing them under the care of the Fisheries and Game Department; they later became part of the national park.[113]

Those who undertook the slaughters, and the regulatory structures that have at times supported these activities (such as the bounty), ultimately sought to achieve an immediate, drastic reduction in pelican numbers or local extinction. While pelican numbers may have declined in the area because of habitat changes and slaughters, we can also see their continued nesting on the islands as a defiance of these attempts, whereas breeding on Kangaroo Island ceased. Reid has called pelicans "flexible responders," not only for their ability to take advantage of sudden floods but also because they have found new food sources, including rubbish dumps, and nesting sites in changing landscapes, such as artificial islands.[114] Likewise, their continued nesting on the Coorong demonstrates their ability to adapt to new circumstances. Yet overall their numbers have declined in the Coorong region, which ecologists attribute to reduced freshwater inflow due to diversion for agricultural uses, a trend that has been exacerbated by climate change. Between 1985 and 2006 their numbers declined by 85 percent, with sharp declines also recorded for other bird species in the area.[115]

Contemporary problems pose new questions, but the complex, conflicting, and contested values and what is at stake in the outcomes resonate with those of the pelican slaughter of 1911. Indeed, contemporary debates in Australia about culling kangaroos and dingoes to limit the damage they cause on farms have strong correlations with the issues discussed in this chapter, and in some ways cannot be fully understood without this history. In both cases, Aboriginal people, famers, conservations, biologists, and humanities and social science scholars have debated whether killing these animals is effective or moral, with conflicting ideas about death, protection, and what is allowable on common and private property.[116]

We continue to live in contested landscapes and with the legacies of these past disputes. The slaughter and its fallout reveal the long legacies of bird and fauna protection, the persecution of particular species, as well as the deep roots of more recent conflicts over the exclusion of people from government and private protected areas. Further, long-running class conflicts and entrenched racist discourses came to bear on the regulations that related to the slaughter and influenced its fallout, including the legal exclusion of some groups from the islands. We continue to live in landscapes with contested values and with the legacies of the past disputes.

SIX

MIGRATING

Wetlands, Transcontinental Bird Movements, and Global Environmental Crisis

THE WIND WHIPPED UP SMALL WAVES IN THE BRACKISH LAGOON. The warm day had turned cool, and the birds huddled into their ruffled feathers. I was standing on the edge of Fivebough Swamp, a small wetland in southern New South Wales. I could see the many birds on the water and toward the opposite shore, including what looked like a small group of Latham's snipe. One, then another of these birds took flight. Soon the whole group was in the air, flying northward along a low mountain range. I was lucky to see them, as these birds are not frequently sighted, often expertly camouflaging themselves in small sedges and bushes. These birds breed in Japan and visit Australia during the Southern Hemisphere's summer months. Along with being listed in the Ramsar Convention on Wetlands of International Importance jointly with nearby Tuckerbil Swamp since 2002, this swamp was also subject to a treaty with Japan on birds that migrated between the two countries.[1] Indeed, these birds were part of the reason this swamp could be called a wetland at all.

Engaging with Australian federal government archives from the early 1970s, this chapter examines a fairly narrow but important episode in this history of wetlands, focusing on a key period when "wetlands" was new terminology in the language of conservation. While I am more used to writing about archives grounded in particular watery places, as with the other chapters in this book, I have come to appreciate that some of the critical sites in a history of wetlands are boardrooms and government buildings. Further, the specific wetlands informing this book, and the many other wetlands in the Murray-Darling Basin, have been carried along and shaped not only by decisions made elsewhere but by decisions that might primarily be concerned with other places and agendas. Actors at multiple scales and in various sites have influenced this history of wetlands, including international organizations, the countries with which Australia formed agreements, the Australian government, Australian state and territory governments, scientific organizations, and specific wetland sites, all of which rubbed up against each other. All ultimately contributed to the emergence of a contested category of wetlands, which frequently torqued against the particularities and lived experiences of wetlands as human and more-than-human places.[2] The decisions and disagreements of bureaucrats and scientists in Australia and elsewhere in this period about what counted as a wetland and why have had long-lasting and mixed consequences, including for wetlands in the Murray-Darling Basin.

The wetlands category is historical. It has been shaped by specific values and has helped cocreate particular kinds of imagined ecologies. In his study of the forests of Canada's west coast, Bruce Braun has argued that "something called the forest," far from being "self-evident... and in an space outside of politics,... *enters history.*" Like forests, wetlands have been "made visible."[3] The term "wetlands" was used before the 1960s, but it was not common. In the 1960s and 1970s, wetlands had a historical moment; they entered history internationally as a category and an object of conservation.[4] "Wetlands" belongs to new globalizing terminology that emerged after World War II—which includes "the environment"—to describe growing concern over negative human impacts on the world.[5] This chapter focuses on two key factors that together shaped the emergence of wetlands as an object of conservation in the 1960s and 1970s and are important in a history of wetlands more generally: bird-centrism and multiscalar politics. Each has had particular stakes, creating lasting tensions in wetlands conservation and management.

In Australia, bird-centrism was embedded in the category of wetlands and in the knowledge informing wetlands conservation and management, becoming mutually reinforcing with international agreements. Other kinds of biota, like mosquitoes, have clearly influenced aspects of wetlands management, including their drainage and spraying. However, in the 1960s and 1970s, wetlands gathered meaning as valuable primarily as bird habitat. The wetlands category took shape within particular knowledge practices, namely those of the sciences. This category was also shaped by gaps in knowledge within these practices, including the values and knowledge of Aboriginal people as well as the migratory status and routes of some birds. This chapter, and the book as a whole, aim to contribute to what Braun has called a "critical environmentalism."[6] Rather than undermining critiques of swamp drainage and modernist water management, it aims to situate wetlands within a longer and broader set of cultural practices and power relationships.

Along with bird-centrism, multiscalar politics was critical in the emergence of wetlands as a category and an object of conservation. For instance, international agreements on wetlands and migratory bird conservation have deeply influenced how this category has taken shape. These agreements were shaped in turn by particular national politics and touched down in and were remade by local places. These dynamics influenced how the wetlands category was taken up in Australia and elsewhere. Crucially, "wetlands" provided an international language that grouped and renamed a variety of watery places, smoothing over local variations in terminology and allowing for an international agreement on these places, namely the Ramsar Convention on Wetlands of International Importance, established in 1971. In Australia, a national environmental movement, Pacific diplomacy, and scientists' concerns over species and habitat loss, which were shared more widely, converged to shape the Australian government's involvement in the Ramsar Convention and simultaneously a Japan-Australia Migratory Bird Agreement. These two international agreements need to be understood as connected, as the Australian government sought to shore up the conservation of both migratory birds and their watery habit, at the same time enhancing its own influence over conservation matters in the states. Government conservation departments leveraged the agreements jointly to initiate a nationwide wetlands survey, which was ultimately abandoned for multiple reasons, including disagreements over whether wetlands were an object of conservation primarily as waterbird habitat or for other uses and values.

The movements of transcontinental migratory birds, such as Latham's snipe, are central to this history and help us think about the many, sometimes distant connections among watery places. Birds have shaped diverse uses and values of wetlands at various scales and connected places through their bodies as they move between sites.[7] Moreover, a focus on nonhuman agency can help us think in new ways about the diverse and wide-ranging biocultural relationships, networks, and more-than-human histories that have shaped places and their connections, illuminating the sometimes profound consequences of how these have been valued and understood.

THE AUSTRALIAN ENVIRONMENTAL MOVEMENT, BIRD MIGRATION, AND GOVERNMENT CONSERVATION

Australian representatives had not been involved in the conception or drafting of the Ramsar Convention, the text of which was finalized in the Iranian city of Ramsar in 1971. Yet Australia became one of the first nations to sign the convention in 1972 and was the first full member nation in 1974.[8] Moreover, Japan, rather than Australia, initiated a migratory bird agreement between the two countries in 1972, yet Australian officials eagerly accepted the proposal, signing the final text in 1974.[9] The involvement of the Australian government in both the Ramsar Convention and the Japan agreement needs to be understood in the context of a growing national environmental movement, which reacted against the environmental effects of postwar and Cold War national development projects, including intensified land settlement supported by more dams and irrigation, changes in industrialized farming, increased wetlands drainage, and the severance of floodplains from rivers by extensive mitigation measures.[10]

In this context a national election saw a new (left) Labor government led by Gough Whitlam elected in December 1972. The Whitlam government soon began to seek greater involvement for the Australian government in environmental matters and created a new Department of the Environment and Conservation.[11] Just two months before the Whitlam government was elected, the Australian House of Representatives released a committee report titled *Wildlife Conservation*. The committee, established in 1970, was made up of representatives from both major political parties. In the report, the committee members wrote that "perhaps the most significant fact to emerge from the Inquiry was the general lack of knowledge about wildlife distribution and ecology" in Australia, and they recommended that a national wildlife survey be undertaken.[12] They further found that "the major cause of decline [in animal population numbers] is undoubtedly the change in the habitat of the species concerned, bought about principally by land clearing and changes caused by grazing" rather than hunting.[13] These statements resonated with the growing concerns of biologists since the 1950s that habitat loss was a significant issue for many species, especially birds.

This report incorporated a review of the need to protect migratory birds. Of the sixty-six species of birds listed as transequatorial migrants, the committee heard evidence that only one needed an international agreement for its protection: the Latham's snipe.[14] Harry Frith, then the chief of the Wildlife Division of the Commonwealth Scientific and Industrial Research Organisation (CSIRO), was one of the key people to give evidence to the committee. It is likely that he brought the birds to the committee's attention, as in 1970 he had conducted a research project on them in collaboration with Japanese researchers. Japanese researchers were becoming concerned over the pressures of increased industrialization and urbanization on parts of the birds' breeding habitat, and soon afterward Yoshimaro Yamashina, a prominent Japanese ornithologist, included it as one of the endangered birds of Japan.[15] Frith and his Australian colleagues published their findings on Latham's snipe in 1977, emphasizing that a loss of habitat due to intensive land use, wetland drainage, and the damming and canalization of rivers as well as hunting had potentially altered the distribution and reduced the populations of the birds in eastern Australia. NSW had only recently banned hunting of the birds, and Victoria, Tasmania, and Queensland (Latham's snipe was a passage migrant in Queensland) all allowed hunting during key migration times, when the birds were relatively vulnerable. Frith and his team stressed that so little scientific research had been undertaken on the birds in Australia that making any definitive claims, even of distribution, was difficult.[16]

The committee members viewed international agreements for bird protection as a politically astute move, writing that they "regard it as part of Australia's international responsibility to ensure not only that action is being taken in this country but also to demonstrate our concern internationally through the establishment of agreements with other countries." In addition, they argued that species protection needed greater alignment between Australian states and territories, as different protection and hunting laws prevailed in each. Ultimately the committee recommended that Australia "seek unilateral agreements with the Governments of Papua New Guinea, New Zealand and Japan" to protect migratory and, in the case of the first two, also nonmigratory birds that were common to these countries.[17]

In a section on wetlands and swamps, the report stated, "It is generally agreed that waterfowl are declining in numbers as a result of the encroachment of agriculture and resultant drainage of swamp land, the damming of rivers, flood mitigation programmes and the trampling by stock and feral animals of nesting cover on the edge of lagoons." It recommended that those undertaking land drainage and water storage schemes consider effects on waterbirds and their breeding grounds on the basis of hunting and aesthetic value. The committee saw this as important in Australia because it was "the poorest continent in relation to the number of species and the size of populations of waterfowl that it supports," having just "nineteen [game] species." It is notable that the report discussed "waterfowl," even then seen as an old-fashioned term that reflected hunting interests and referred specifically to game species. While the committee acknowledged that "many wildlife conservationists regard protection as the main obligation to native fauna and cannot accept the idea of conserving for hunting," it continued to justify protection on the grounds of waterbirds' value as game or resources, and as holding aesthetic value, all of which have long histories in Australia and draw on British traditions of protection.[18] In the Australian *Wildlife Conservation* report, we can nevertheless see a considerable weakening of the alliance between hunters on the one hand and ornithologists and humanitarian advocates on the other in the context of a growing environmental movement. This movement was emerging in different registers, including a dynamic reciprocity between local, national, and international scales. Notably, the report used the new, international conservation term "wetlands."

THE RAMSAR CONVENTION, BIRD MIGRATION, AND THE GLOBAL ENVIRONMENTAL CRISIS

The rise and popularization of wetlands as a category and object of conservation around the world was significantly influenced by a series of international discussions and research projects on migratory waterbird conservation from the early 1960s, which resulted in the Ramsar Convention. The meetings were led by waterbird protection organizations in Britain and Europe, including the International Wildfowl Research Bureau and the International Council for Bird Protection, which were concerned about declining numbers of migratory waterbirds and sought international cooperation for their protection and conservation of their habitats. The focus of the planned convention became wetlands habitat rather than migratory birds in 1965, largely due to a sense among participating countries that it would be too slow and difficult to secure agreement and ratification from so many countries on a migratory bird convention, and agreement on habitat protection would be easier.[19] Historian Robert Boardman has argued that "wetlands were for the 1960s what arid regions were for the 1950s," as they provided a focus for diverse international interests and facilitated cooperation between countries.[20] However, the foundational defining common interest was migratory waterbirds. As Boardman has written elsewhere, "Of the conventions of the period, this came the closest to being a bird convention."[21] This bias toward birds remained in the Ramsar Convention and is still evident today. In fact, the original title, "Wetlands of International Importance *Especially as Waterfowl Habitat*," remains unaltered in the latest version of the agreement, although this last part has been dropped everywhere else in Ramsar's public profile.[22] This bias also remains in wetlands management and conservation, meaning that other, less charismatic species have often been overlooked.[23] However, that the Ramsar Convention became a *habitat* convention has meant that it has been used for many other purposes, like the protection of crocodiles in parts of Australia and South Africa.[24]

The Ramsar Convention emerged in the context of a growing global sense of environmental crisis: a widespread recognition that many issues, from the protection of migratory birds to nuclear waste and air pollution, needed global cooperation as these issues crossed borders.[25] Biologist Michael Soulé has argued that conservation science, which championed biodiversity, emerged in this era as a crisis discipline responding to the sense that there needed to be immediate action on pressing problems of species loss and environmental destruction.[26] So too did the conservation science subfield of wetlands ecology, which emerged out of a sense of crisis in the reduction of species-rich watery places, linked to shrinking numbers of waterbirds, something that had been noted in the United States and Europe for several decades.[27]

The term "wetlands" became useful in discussing these places and attempting to conserve them at an international scale. The term had been used in the United States from the turn of the twentieth century as a synonym for "swamps." By the 1950s, American scientists had begun to use "wetlands" in a conservation framework to refer to a diverse range of watery places, and they had begun to gather regulatory definitions in the United

States in the 1960s and 1970s.[28] In the 1960s it also became an international lingua franca for scientists, replacing local terms like "swamps," "marshes," and in Australia, "billabongs." In addition to providing international terminology, the wetlands category served to repackage watery places—commonly associated in the West with disease, death, waste, and the underworld—as places worth protecting, particularly as migratory bird habitat.[29] Wetlands were therefore constituted with their reevaluation.

The Ramsar Convention took the broadest possible definition of wetlands so as to encompass local diversity in how this term might be applied: "Areas of marsh, fen, peatland or water, whether natural or artificial, permanent or temporary, with water that is static or flowing, fresh, brackish or salt, including areas of marine water the depth of which at low tide does not exceed six metres."[30] Using this definition, coral reefs could be included as wetlands, something that wetland ecologists still take issue with. The Ramsar definition remains contested.[31]

The Ramsar Convention has become known as "the first modern environmental convention," because it aimed to protect habitat rather than regulate hunting and reflected the ethos of the emergent environmental movement. It showed a new global recognition that hunting was no longer always the biggest threat to animal populations, that reductions in habitat were often more significant, and that global cooperation was necessary for wetlands protection.[32] Yet Ramsar's focus on waterbirds, and its processes for protection—essentially the establishment of a list of internationally important sites by participating countries, which were obliged to create protected areas—can also be seen as carrying forward the interests of ornithologists and hunters who had traditionally dominated the protection of these birds and their habitats in places like the UK, Europe, the USSR, the United States, and Australia. Indeed, the UK, USSR, and European countries led and dominated the Ramsar Convention discussions.[33]

Despite its emphasis on the "wise use" of wetlands by people, the Ramsar Convention's focus on establishing protected areas helped to perpetuate the colonizing tendencies of this mode of environmental protection in settler societies like Australia and the United States. By the 1970s, Australian government and conservationist agendas for protected areas were heavily influenced by an idea of wilderness as devoid of human activity and history, which erased the past and ongoing connections of Aboriginal people.[34] These kinds of protected areas continued earlier exclusions of Aboriginal people from particular sites. At the same time, an Aboriginal land rights movement in the 1960s and 1970s created opportunities in some places—particularly the Northern Territory—to establish protected areas comanaged by Aboriginal groups and the government starting in the late 1970s.[35]

It is important to note that the wilderness approach of the government conservation movement, which centered on protecting remnant natural places from people and purging them of human uses, also excluded other people such as white working-class residents living along and using rivers and wetlands for various purposes. Indeed, while wilderness agendas were most prominent in this period, there were other views of protected areas in Australia at this time. This included the view that national parks should be used and changed by people—that is, be places for people—often championed by the working class and farmers, an idea with roots in the late nineteenth century. These differing views can be broadly understood as cutting along class lines, with a wilderness rationale more often associated with middle-class conservationist arguments.[36]

Local circumstances and histories of area protection were important in how the Ramsar Convention touched down in different places, itself shaped by concerns in specific places, especially the UK, Europe, and the USSR. As various countries signed and ratified the Ramsar Convention, wetlands were made visible as objects of conservation in new ways, with specific legal, bureaucratic, economic, and political associations; these were in turn reconfigured by local ecologies, histories, expertise, politics, and ways of life.

AUSTRALIA: RAMSAR AND THE JAPAN-AUSTRALIA MIGRATORY BIRD AGREEMENT

Soon after the creation of the new Department of the Environment and Conservation in the Australian government, the minister, "Moss" Cass, started the process of Australia becoming a full party to the Ramsar Convention; he also wrote to the International Union for Conservation of Nature requesting information on other international agreements to which Australia could become party.[37] The Whitlam government employed a strategy of shoring up federal involvement in conservation through greater involvement in international environmental conventions and treaties. Part of this strategy involved strengthening its ability to use its external affairs power. This power, enshrined in the Australian constitution, allowed the federal government to become involved in state matters if these lay within the terms of international agreements.[38] In the early 1970s, this strategy became so prominent in the Whitlam government that the attorney-general warned that reliance on it could have significant political and legal ramifications affecting the functioning of the government.[39]

Australia's participation in both Ramsar and a migratory bird treaty with Japan was linked to this strategy. For example, an internal report from the Department of the Environment and Conservation stated that "the main advantage of the Australian Government becoming a Contracting Party [to the Ramsar Convention] is that under the external affairs powers of the Constitution, it could technically assume jurisdiction over wetlands covered by the Convention." Others, following the advice of the attorney-general, viewed this power as difficult to enforce without declaring wetlands to be "national parks and nature reserves." The practical use of the external affairs power remained contested in these years but came to a head in the early 1980s when a Labor federal government used legislation that ratified the World Heritage Convention to stop the Tasmanian Liberal (conservative) state government from damming the Franklin River. This intervention by the federal government was upheld by the Australian High Court in a landmark legal case and was widely seen as a victory for the national environmental movement.[40]

Australia's obligations to the Ramsar Convention and the migratory bird agreement with Japan could be met under a new National Parks and Wildlife Commission and Service, officially created in 1975.[41] The coordination and extension of a national protected areas system through this agency, which aimed to work with similar state agencies, was another key strategy of the Whitlam government to facilitate greater federal involvement in environmental conservation, fulfilling its obligations to multiple conventions and treaties.[42] International diplomatic discussions consolidated approaches to wetlands as objects of conservation around the world. Through these discussions, processes, and resultant agreements, wetlands in Australia became associated and overlaid with protected area systems at various scales. Wetlands therefore took on many of the meanings given to protected areas by Australian government and conservation agendas, which were embedded in colonial ideas of wilderness and a separate nature that excluded humans.

Scientific experts contributed alongside bureaucrats to Australian negotiations for both the Ramsar Convention and the Japan agreement. Harry Frith, chief of the Wildlife Division of CSIRO, was one of the main scientific experts who influenced Australia's involvement with the international agreements. Frith's ongoing research and interests in bird biology and behavior shaped the way Australia engaged with these agreements. Notably, he reinforced the bias toward birds evident in the Ramsar Convention and helped facilitate the bird-focused agreement with Japan. Since joining CSIRO in 1946 at the age of twenty-five, Frith's research had largely focused on birds.[43] In line with government priorities, Frith often undertook applied research that sought to limit damage to crops from pest birds, including ducks in the rice-growing areas of the Murrumbidgee Irrigation Area in NSW and magpie geese in the rice-growing areas of the Northern Territory in the 1950s. These studies had often led to life histories of the birds and recommendations that farmers adjust their growing methods rather than employing pest control measures that were often harmful to birds. In the case of magpie geese, his studies revealed that their numbers were declining and that rather than being pests, they were in need of protection. This work helped support the creation of Kakadu National Park in 1979. He had also started collaborating with Japanese researchers over the protection of Latham's snipe, which migrated between the two countries.[44]

At an interdepartmental meeting on the Ramsar Convention in 1973, Frith commented on the vast number of important wetlands in Australia. He noted that these comprised "swamps, coastal salt lagoons and estuaries" that were "important as habitats for many birds which carry out an annual migration from Asia to Australia. Of about 69 species of birds involved in these migrations approximately 50 are dependent on wetland habitats."[45] In such statements, Frith drew strong connections between the conservation of wetlands and migratory birds, which resonated with the agreements and international trends, and he perhaps leaned into the practical opportunities these offered for conservation.

Frith worked within the political constraints created by Australia's involvement with the Ramsar Convention. To become a full party to the Ramsar Convention, Australia needed to nominate at least one wetland with acceptable protections to list as internationally important. Despite the number of important wetlands in Australia, nominating an initial wetland in the states for Ramsar listing could potentially raise political difficulties, including questions about the implications of international environmental conventions for federal intervention in state governance arrangements. Frith suggested nominating Cobourg Peninsula Wildlife Reserve and Sanctuary, located in the Northern Territory and governed directly through federal government legislation, thus avoiding this problem. This wildlife reserve already had the appropriate protections, and there was extensive scientific research to back up the nomination, including Frith's work showing that various kinds of migratory birds visited the mangrove estuaries to breed and to winter. Further, birds that did not migrate nested at the site, including magpie geese (*Anseranas semipalmata*). This became not only Australia's but the world's first Ramsar-listed wetland of international importance.[46]

In 1973 Minister Cass wrote to Prime Minister Whitlam with the proposed nomination, recommending that Australia become a full party to the Ramsar Convention and noting that "swamps, coastal salt lagoons and estuaries in the northern, eastern and southern coasts of both the mainland and Tasmania . . . are not only of Australian significance but are also internationally important as they provide habitats for waterbirds which annually migrate southward from Asian countries. Already Japan has initiated discussions with Australia in an effort to conclude a treaty for the protection of these migratory birds and their environment and I hope that negotiations for similar agreements with other countries in our region will be forthcoming."[47] Japan approached Australia about signing a migratory bird treaty in 1972.[48] Japan sought to sign a number of similar agreements, including with the United States and the USSR, as steps toward possible pan-Pacific migratory bird protection or even a global convention on migratory birds. For all of the countries involved, migratory bird agreements were tools of diplomacy in the Pacific at a time when Cold War tensions were still high. For the Australian government, the agreement with Japan was certainly a foundation for building closer relationships between the two countries. Further, agreements like the Ramsar Convention demonstrated governments' commitment to environmental protection on an international stage. The emergence of a global environmental movement had made the environment a key aspect of international politics and power brokering.[49]

The Pacific migratory bird agreements reflected genuine concerns over the need for greater cooperation in conservation efforts. In fact, some in CSIRO's Wildlife Division urged that an agreement with Japan should not just focus on birds but include a broader set of conservation measures and more comprehensive environmental standards. Yet Japan advocated an agreement on birds, and in both countries there was widespread concern over extinctions and decreases in bird numbers, with collaborations between scientists on particular species such as Latham's snipe already emerging. The agreement aimed to facilitate exchanges in information among scientists as well as the establishment of bird banding programs. It also required a shared commitment from the national governments to protect listed species of birds and their habitats, including through protected areas.[50]

The negotiations revealed to those involved that there had been very little government research on migratory birds in Australia. In the nineteenth century, many colonial ornithologists had focused on collecting eggs as well as establishing the location of breeding places and migratory routes of birds, often between hemispheres. Historian Libby Robin has argued that this was because the bird migrations resonated with their own travel or migration to Australia from Britain and Europe.[51] One of these was Victorian ornithologist and avid egg collector A. J. Campbell, who dedicated significant time to working out where Latham's snipe bred. In a 1893 newspaper article he described a long-held idea that the birds bred in the "unknown far north-west interior" of Australia; colonists had assumed that the birds "probably . . . bred in autumn in countless companies in great marshy areas as yet only explored by the rude savage."[52] Though ornithologists sought to produce objective science, they were embedded in colonizing processes, often ignoring or appropriating Indigenous knowledge and reinforcing colonial ideologies of racial hierarchy.[53] Campbell focused his search for Latham's snipe eggs in Japan after another ornithologist, Henry Seebohm, published his observations of the birds there.[54] In 1898, Campbell could finally claim to have found where the birds bred when his hired egg collector gathered specimens from the birds' nests at the foot of Fujiyama (Mount Fuji), and sent some samples back.[55] The language of discovery of birds and

FIG 6.1 Neville Cayley, *Australian snipe, 1902*, National Library of Australia, http://nla.gov.au/nla.obj-137969698.

their eggs is prevalent in these ornithologists' writing, in many ways resembling writings by colonial explorers.

While the scientists sought to discover birds, Aboriginal people had deep knowledge of them that largely centered on relational observations. Many Aboriginal groups have linked bird movements, including local and more distant migrations, to observations of weather changes. For example, the Yanyuwa people, whose Country is in the Northern Territory, have connected the appearance of coastal rain clouds with the beginning of bird migrations.[56] In the Lower Murray River, Yaraldi people have linked regionalized brolga movements to seasonal changes in wind.[57] For some Aboriginal groups, particular migratory birds are important in kinship ties and obligations: for example, *Koltoli* (Latham's snipe) is important to the kinship obligations of Limpindjeri, who are part of the Yaraldi and whose Country is east of Lake Albert in South Australia.[58] Many Aboriginal groups have linked the behaviors of animals and plants to availability of resources.[59]

Historical scholarship on birds in Australia has largely focused on the challenges faced by ornithologists in moving beyond British and European models of bird behavior, especially annual seasonal migration. Ornithologists only gradually developed an understanding that most Australian birds were influenced more by rainfall than seasons—that is, that they were nomadic rather than migratory.[60] This scholarship has been important in showing the incompatibility of European understandings of animals, plants, climates, and environments with conditions in Australia. But what of the birds that did undertake seasonal intercontinental migrations? These migrations, and how ornithologists understood them, shaped and connected distant places with a range of mixed, and important, political and ecological consequences in Australia (and elsewhere). They can also help to illuminate the diverse and wide-ranging biocultural networks that have connected Australia with the wider East Asian and Asia-Pacific regions and beyond.

By the turn on the twentieth century, ornithologists knew that various birds migrated between Australia and parts of Asia, breeding in the Northern Hemisphere and wintering in Australia.[61] International networks of ornithologists contributed to this understanding, for instance through Russian ornithologists publishing observations in Australian journals.[62] Knowledge of bird movements was, however, far from certain. Ideas about the precise routes taken by birds such as Latham's snipe were speculative and contested, and some species that did not leave Australia were thought to be migratory by both ornithologists and amateur observers.[63]

By the 1860s Latham's snipe were already familiar to waterfowl hunters in southeastern Australia, who prized them for sport as well as food. Snipe required skill and patience in hunters; in fact, the word "sniper" comes from the mode of hunting used to kill snipe. In the early twentieth century people began to record declines in Latham's snipe in southeastern Australia, and hypothesized that it was due to a loss of habitat from wetland drainage, intensified land use, and overhunting.[64] According to later government estimates, hunters killed up to ten thousand birds annually in Australia in this period, including many in Victoria and Tasmania.[65] In the late nineteenth and early twentieth centuries, concerns that birds were being overhunted for a variety of reasons—including for plumes, as agricultural pests, and for meat—prompted advocacy groups to actively lobby governments for greater protection of the birds.

These groups often directed their efforts toward protecting native birds that were useful (because, for example, they ate insect pests on farms) or aesthetically pleasing. Seen as not particularly pesky, useful, or beautiful, nor truly native, Latham's snipe and many other transequatorial migratory birds seem to have been marginal in these debates. Ornithologists, advocacy groups, and hunters also aimed to protect breeding sites rather than habitat, and many wetlands were protected as game reserves in this period. But because Latham's snipe and other transequatorial migrants bred in the Northern Hemisphere, they were not included in this either. Bird protection advocates, including sport hunters, argued that birds were vulnerable and in need of protection during breeding, and at other times it was, broadly speaking, expected that they could go almost anywhere. Further, while hunters viewed Latham's snipe as good sport, it was not a key game species. Because no migratory birds were significant game species, Australia did not seek international agreements to protect them in late nineteenth and early twentieth centuries, as happened in North America.[66]

In the first half of the twentieth century, government ornithological research focused largely on species that farmers and fishermen viewed as pests, like cockatoos that raided orchards, ducks that were blamed for reducing rice crops, and pelicans that ate marketable fish. It was the research in this period that provided scientific evidence for nomadic and opportunistic behavior in many Australian birds, behavior that was already generally understood. Robin has noted that in the 1920s and 1930s this research on nomadism was influenced by international interest in "the physiology of irregular breeding" in birds, including the influence of rainfall.[67] Following an intense dry period and sand drift in the inland in the 1930s and 1940s, Australian biologists turned more directly to studying desert or arid zone birds in the 1940s and 1950s.[68] In broad terms, this extended research on nomadism and opportunism. Australia only established a government-funded national bird-banding program in the 1950s, something that the United States and Britain had started much earlier. As Robin has noted, "Migration has traditionally been the major interest of banding studies," and this interest had not been significant in Australian government research.[69] In addition to migratory birds falling outside the purview of applied government science, the lack of research on them in Australia in this period might be further explained by the government's focus on national projects and development during and following two world wars that had created turmoil and sensitive international relations in the Asia-Pacific region.

Through most of the twentieth century, migratory birds with routes into East Asia and the Pacific seem to have gained significant popular and scientific attention only as potential carriers of disease. These views became part of the contested biocultural terrain of immigration laws that discriminated against nonwhite migrants, including from these regions.[70] The first half of the twentieth century was a period of heightened racism in Australia, which continued in the Cold War period. A largely conservative Australian population saw links with Asia, including through birds, as undesirable. In the early 1950s, in the wake of a significant outbreak of encephalitis in the southern Australian states, medical professionals claimed that migratory birds had introduced Japanese encephalitis to Australia, which mosquitoes then spread to humans. Newspaper columnists interpreted the introduction of the disease in the context of racist ideas of a clean, white Australia located in a diseased, nonwhite region. One stated, "Research points to migrating birds, travelling to Asia by way of the Pacific Islands north of Australia [and back], as carriers of the disease to our mainland. If this is so, it is quite possible that the disease first appeared in the islands, or in Asia rather than in white Australia."[71] Another stated that "Murray Valley encephalitis, which broke out in 1951 gave the Murray valley a bad name. But the disease is now known to be Japanese encephalitis possibly carried to Australia by migrating birds."[72] The kind of encephalitis that had caused the outbreaks in Australia was, however, later shown to be a slightly different strain that was local to parts of Australia.

Studying the role of migratory birds in spreading Japanese encephalitis in East and Southeast Asia was one of the key rationales behind the establishment of the Migratory Animal Pathological Survey in the 1960s. This effort was funded by the US Army and the Southeast Asia Treaty Organization and led by American ornithologist Elliott McClure. The study included over fifteen countries and involved the banding of thousands of birds. Through this research McClure devised the idea of the East Asian Flyway as a major bird migration route, which included Australia (later subsumed into the bigger East Asian–Australasian Flyway). The flyway concept had been developed by researchers in the United States in the 1920s and 1930s and had been a major influence on wetland management along migratory paths in North America. From this research, McClure argued that changes in bird habitat in Asia and Australia had altered the migration routes.[73] It was perhaps a result of this work that Australian researchers led by Frith had begun studying Latham's snipe in New South Wales in the 1970s.

Relatively little scientific research had been undertaken on migratory birds in Australia, and at the time the agreement with Japan was signed in 1974 the migratory status and routes of many birds were still contested. A starting point of the negotiations between Japan and Australia toward an agreement was the compilation of lists of migratory birds by each country that would be subject to the agreement. With a great deal of uncertainty over the migration status and routes of many birds, the countries started with lists of birds present in both countries. When Australia and Japan submitted lists of birds to include in the annex to the agreement, Japanese researchers questioned the evidence for several birds on the Australian list, which had been recorded through sight identification. The Japanese government insisted that any bird listed needed documentary evidence in the form of a photograph or museum collection (for example, specimens or eggs).[74] Further, Australian and Japanese ornithologists used different nomenclature and in some instances different taxonomic classifications, which meant that birds might be classified as belonging to one subspecies in one country and another subspecies in the other, thus affecting whether they were listed as common to both countries. The two countries took a pragmatic view, opting for the simplest list in order not to stall negotiations, with the view that more bird species and subspecies could be added later. The agreement ultimately covered "migratory birds *and* birds in danger of extinction and their environment," partly due to disputes over the evidence for migratory status and routes of birds.[75]

The compilation of these lists raised the question of whether Papua New Guinea could or should be included in the agreement, as the scientists advising on the agreement were concerned that some birds migrated between Papua New Guinea and Japan but did not come as far south as Australia. At the time, Papua New Guinea was an Australian dependent territory, and the Australians saw its inclusion as important for bird conservation. Yet the future national independence of Papua New Guinea was an important and highly political issue that factored into these negotiations. In response to an Australian request to include Papua New Guinea in the agreement, the assistant secretary for law in the Territory of Papua New Guinea, G. P. M. Dabb, pointed out that there would be "certain difficulties ... with the view to Papua New Guinea becoming party to the agreement in its own right after independence." The agreement had been initiated as a bilateral treaty that primarily concerned birds migrating between Japan and Australia; in his view it "did not lend itself to very well a multi-lateral situation." Further, he noted that most of the migrating birds would stop in both Papua New Guinea and Australia, and thus would be covered by the agreement in any case. He further wrote, "I would not imagine that there would be any reason why Papua New Guinea would not be willing to extend protection to these particular birds [when they stopped in Papua New Guinea]. However, as a matter of policy, conservation being a relatively western and sophisticated concept, it might be more appropriate for Papua New Guinea not to enter into a specific agreement at the insistence of Australia or Japan but to wait and see what sort of co-operation might develop among other Pacific Island countries of the region."[76] Here it is evident that Papua New Guinea wanted to clearly separate itself culturally and politically from Australia, and looked more toward the Pacific Islands for biocultural synergies. The possibility of naming Papua New Guinea in the agreement raised another issue: Japan was concerned that naming territories in the agreement with Australia could disrupt its agreement with the USSR, as this could raise the issue of territories disputed between Japan and the USSR. Ultimately, Papua New Guinea was not named in the agreement.

In the initial drafts of the agreement, there were provisions for scientific collection of birds and their eggs, as well as for annual open seasons for hunting, but there was limited explicit consideration of the effects of the conservation measures on the uses and values of migratory birds to Indigenous people in either Japan or Australia, nor of the effects of establishing sanctuaries and protected areas to conserve bird habitats like wetlands, which could limit Indigenous people's access to their traditional lands, spiritual sites, and resources. In an era of Aboriginal land rights in Australia, these problems were spotted early on by Commonwealth officials working on the agreement. However, accounting for this within the narrow confines of conservation ideals informed by a colonizing concept of wilderness, often blind to the long histories of Aboriginal people in Australia, was not straightforward.[77] As the agreement went through several rounds of revision, the Australian representatives appear to have addressed this problem with vague wording—for example, including an exception on bird protection for "specific purposes" through which Aboriginal people's traditional hunting rights might be accounted for. Japan, however, addressed the rights of Indigenous people directly, arguing for inclusion of a clause that protected the subsistence hunting rights of people in specific areas, thereby accounting for traditional hunting by Ainu and other Indigenous people in Japan, who largely practiced these activities in the northern islands.

With provision only for subsistence hunting and egg collecting, where did this leave local, including Indigenous, commercial industries reliant on migratory birds that would come under protection? D. L. Serventy, a recently retired CSIRO Wildlife Division researcher, manager, and mutton bird expert, wrote a letter to Frith raising the issue of how this could affect commercial activities that might not be regarded by government officials as "traditional," such as the mutton bird industry in Tasmania. Aboriginal people, who have long hunted mutton birds, established an industry with European sealers and others that helped sustain a number of communities during the nineteenth and twentieth centuries. Each year the birds' young were hunted during the nesting months on islands in the Bass Strait off the coast of Tasmania, just before they were ready to migrate. This was a commercial industry, and many of the birds killed were sold.[78] Serventy noted that far from being endangered, the birds were "undergoing a population explosion."[79] However, Frith thought the industry could be provided for under other clauses. These specific examples, raised by individuals and noted in the files documenting the negotiations with Japan, reveal many other instances in which Aboriginal people's activities and economies were not considered. In fact, representatives of the Australian Foreign Affairs Department admitted to their counterparts in Japan that "we do not know" about the values and uses to Aboriginal people of the birds listed in the agreement. The Japanese government requested information on the number of Aboriginal people in Australia, and the Foreign Affairs Department supplied census information with the statement that "the majority of these people do not live in areas to which the birds listed in the annex migrate."[80] All of this tied into and pointed toward ongoing histories of colonization, including the forcible removal and exclusion of Aboriginal people from their land.

A WETLANDS SURVEY

For Frith and other government scientists, Australia's involvement in the Ramsar Convention and the migratory bird treaty with Japan further highlighted the paucity of knowledge about what now might be classified as wetlands on a national scale. Individual studies showed that there had been a loss of important waterbird habitat in specific places, such as a 1970 study which indicated that 60 percent of wetlands along coastal NSW had been destroyed or degraded largely due to drainage for flood mitigation.[81] Yet any effort to quantify losses more widely was difficult, perhaps amplified by the fact that the wetlands themselves were only just being made visible as objects of scientific study.

Frith and other members of the Australian Committee on Waterbirds—made up of state and federal government researchers and managers—sought to provide better information on wetlands as waterbird habitat, to support Australia's obligations to both the Ramsar Convention and the Japan agreement. In 1972, they submitted a proposal: "A Survey of Wetland Habitats of Australian Waterbirds."[82] This proposal was approved, but then the Council of Nature Conservation Ministers significantly widened the brief to "go beyond an examination of waterbird habitat" to "encompass all wetland areas so as to be beneficial to a wider section of government agencies" and provide "data needed for the management and conservation of Australian wetlands" more generally. Ultimately three CSIRO research divisions became involved in undertaking essentially separate investigations as to whether such a survey was feasible: Wildlife, Land Use, and Fisheries and Oceanography.[83]

Notably, the Australian government researchers did not use the Ramsar definition of wetlands, as they sought to reflect Australian ecologies and concerns within the international frameworks. The bird-centered proposal had used a definition of wetlands that focused on their role as bird habitat: "Wetlands are areas of swamp, shallow water or water-logged land. The water cover may be permanent or temporary and the areas are usually characterized by vegetation of moist soil or aquatic type. Because of the food and cover provided by this vegetation wetlands provide a major part of the habitat requirements of waterbirds." Dry land and the "deep waters of rivers, lakes, reservoirs, estuary and open sea" were explicitly excluded because they were not seen to be important waterbird habitat. When the brief expanded, the three divisions needed to formulate a boarder definition. They decided that "wetlands include swamps, marshes, wet meadows, billabongs, lakes, estuaries and coastal lagoons, mangrove flats. These may be temporary or permanent. The mainstreams or main channels of rivers are excluded except for the survey of fishes." This definition was contested, with some representatives arguing that any single definition was impossible, as the parameters of what counted as a wetland would always be driven by specific purposes, and indeed specific sets of expertise.[84]

Each of the feasibility studies soon ran into problems. Led by Frith, the Division of Wildlife Research aimed to test methodologies for classifying wetlands according to the needs of waterbirds, assessing faunal diversity or "richness," and assessing populations of waterbirds. It focused on six sites, all in NSW. While they had some success at these sites, this study threw into question the practicality of undertaking a continent-wide survey. The diversity of bird species and their varied and changing habitat needs made implementing a single methodology too difficult. In addition, constraints of budget and people power meant that comprehensive data simply could not be gathered. The division's report concluded that a national survey "might not be the most important step to take next in waterbirds conservation." What was needed was rather "detailed ecological research."[85]

Ephemeral wetlands presented another problem. Frith admitted that while ephemeral wetlands in Australia were important for their opportunistic use by waterbirds, "no one has yet been able to properly assess them and in our preliminary planning for the National Wetlands Survey Ertz Imagery seem to have very grave deficiencies. At present we have no idea how we will overcome that problem when the survey begins."[86] Ertz imagery came from satellite photographs taken in the 1960s. Frith noted deficiencies in this imagery, including that ephemeral wetlands by their very nature could not be captured only using this data, and cloud cover obscured parts of the continent when the images were taken. Dynamic wetlands in an arid continent proved a challenge for any simple process of quantification. While bureaucratic processes to simplify and make landscapes legible had often served as an important tool in government and commercial exploitation of natural resources, here efforts by Frith and others to make wetlands visible were intended to assist in their conservation.[87] But even for this purpose, and with the single interest of documenting waterbird habitat, they defied any easy measurement.

The other divisions ran into similar problems. For example, with available funding and personnel, the Land Use Division was only able to review existing documentary sources on two previous wetlands surveys in the United States and prior wetland studies in Australia. It argued that these were either inapplicable to an Australian context or not useful when expanded to a national scope, and many had significant issues embedded in their original conception.[88] Further, the report noted that a "continental-scale inventory of Australian wetlands...must be based on a dynamic rather than static description of wetlands. This presents novel problems not tackled in previous land surveys."[89] In addition, the CSIRO divisions pulled the wetlands survey in different directions, toward three different models: wetlands for birds, wetlands as hydrological entities, and wetlands as fisheries and estuaries.

The three divisions, each seeing major obstacles to beginning a national wetlands survey, requested more funds and time for pilot studies in 1976, which would then inform a wetlands survey proper with an estimated cost of $3.3 million over eight years.[90] No more funding was granted, and the wetlands survey was labeled "not essential" by the Australian government, then in the process of cutting spending as the nation faced economic recession. This was done under a new (conservative) Liberal government. Toward the end of 1975, the governor-general, Sir John Kerr, dismissed Prime Minister Whitlam in a historic decision and installed a caretaker Liberal government led by Malcolm Fraser, which won the national election the following month. In 1979, the acting minister of environment and science stated that the wetlands survey was "not implemented because of cost, lack of agreement on a national approach and differences of opinion on the extent to which a national survey should concentrate on the aquatic fauna or the total wetlands ecosystem."[91] The survey had ultimately became unworkable.

That the survey did not, or could not, go ahead has had multiple implications. Perhaps most significantly, wetland ecologists today have limited ability to give robust estimates of losses of wetlands.[92] As recently noted by wetland ecologists, an "understanding of the distribution and extent of wetlands is critical for effective management and policy but remains a knowledge barrier across Australia."[93] The Land Use Division did proceed with a continental mapping of wetlands, but it was incomplete and unpublished, so wetland ecologists have found it to be of limited use. Some regional descriptive information accompanied this map, but it is applicable only at a local scale.[94] Currently, information on wetlands is compiled from patchy state and territory summaries, with only two states undertaking "reasonable" monitoring but using different methodologies. This means that there is "no contemporary national map of wetlands distribution and extent."[95] The "only consistent national wetlands data base" remains a narrow data set from the 1960s that is "plagued" by numerous problems, including inconsistent attribution of types of wetlands and inability to account for ephemeral wetlands.[96] Floodplain and ephemeral wetlands have often been sidelined in conservation efforts in Australia and around the world, often not fitting easily into regular monitoring processes.[97] Instead, a case-by-case and typology approach to wetlands conservation has unfolded, focusing on important or iconic sites that have reasonable historical research behind them.[98]

These areas, or parts of them, have been incorporated into national parks, such as the inclusion of parts of the coastal wetlands of the Alligator River into Kakadu National Park.[99] The Ramsar Convention and the Japan agreement have been extensively used in conjunction with other international agreements—including migratory bird agreements with China (1986) and the Republic of Korea (2007)—in wetland protection and management in Australia. Of the more than 120 bird species found in the Murray-Darling Basin, in 2017 a total of 25 were listed in these international agreements for protection.[100] In 2019 Australia had 65 Ramsar-listed wetlands, with 16 of these in the Murray-Darling Basin. The first Ramsar sites in the basin were four wetlands on the Murray River (Barmah Forest, Gunbower Forest, Hattah-Kulkune Lakes, and Kerang Wetlands), all listed in 1982.[101] As waterbird numbers have declined in the basin—by more than 70 percent between 1980 and 2017—these agreements have on occasion been invoked to prevent wetland drainage and development.[102] The agreement with Australia has been less significant in Japan, which has placed greater political importance on supporting migratory bird treaties with other countries in Asia and has prioritized regional conservation of cranes.[103] Nevertheless, in both places, the agreements provided a new tool for intervening in wetlands and opened up new political possibilities that could mobilize bird migrations to protect particular habitat. In other words, the international journeys made by the birds changed the nature of these places politically. More recently, concerns over H5N1 virus have produced another political shift. People have again become concerned that the long journeys made by birds are also possible routes for pathogens that harm humans, at the same time as immigration debates have flared.

With the ongoing ecological decline of rivers and wetlands in the Murray-Darling Basin, the federal government sought to institute reforms in water management by passing the Water Act 2007. As noted by Jamie Pittock, Max Finlayson, Alex Gardner, and Clare McKay, "The constitutional mandate for this legislation relies in large part on the Commonwealth government's external affairs power to legislate to implement treaties, including the Convention on Biological Diversity and Ramsar Convention on Wetlands."[104] In 2010 the external affairs power for the Ramsar Convention was written into legislation, shoring up this power so that the federal government can more easily and confidently use the Ramsar Convention to intervene in state matters, including the federal government's constitutional mandate for the Water Act 2007. Yet, as others have noted, the potential of the external affairs power to support significant water management reform is limited if the amounts of water needed for environmental and cultural flows are greater than the amount provided for in the specific Ramsar designations of wetlands.[105] Further, the politics of the current (conservative) Liberal government have meant that environmental, and specifically wetlands, conservation has not been prioritized in recent years. In general terms, the importance of the Ramsar Convention for wetlands management in Australia has declined in recent years, as has the potential for significant water management reform under the Water Act 2007. Because the principles and guidelines of the Ramsar Convention cannot be enforced on member nations at an international level, its potential relevance and effectiveness in wetlands conservation is subject to national politics.[106] The Ramsar Convention also fails to address numerous issues raised by the effects of climate change, primarily processes for maintaining the ecological character of wetlands as they adapt to unavoidable alterations and the representation of novel ecosystems arising from climate change. Currently, the convention agreement does not directly address climate change.[107]

Wetland ecologists have continued to criticize the Ramsar definition of wetlands for being so broad as to include any water body less than six meters in depth, so not only reefs—which are more marine systems—are included, but also "all areas of rice cultivation would technically qualify as wetlands, though most such areas are of scarcely any conservation value."[108] Yet since biota, especially much-loved birds, find refuges in areas that people find difficult to see as nature—such as the endangered Australasian bitterns who nest in rice fields—do our understandings of wetlands need to change? Even as these rice fields divert water from wetlands, how might these areas now hold conservation value, and how might our approach to conservation need to change? At the same time, this approach runs the risk of more water being diverted to rice farms from wetlands, with consequences for many species.

In many cases the importance of wetlands for migratory birds in the Murray-Darling Basin has meant that many became subject to the Japan-Australia Migratory Bird Agreement. This includes Fivebough and Tuckerbil Swamps, located among the rice fields of the Murrumbidgee Irrigation Area. Indeed, birds have remained central to wetlands conservation, management, and sciences partly because some are subject to international agreements. For example, all five criteria in Fivebough and Tuckerbil Swamps' Ramsar listing relate to birds, with some of these birds, such as Latham's snipe, also being subject to bilateral international agreements. Management of the wetlands—which includes flood mitigation and sewage treatment—is significantly led by their role as bird habitat.[109] More widely, the health of wetlands in eastern Australia is reviewed each year by wetland ecologists through an aerial bird survey, which also monitors numbers of threatened and migratory birds.[110] Using this information and other sources of data, wetland ecologists have been able to assess the losses of wetlands in the Murray-Darling Basin as the worst in the nation, due mostly to water resource development and management.[111]

While birds are indictors of the general condition of wetlands, the view of wetlands produced by these surveys focuses on birds alone and not other biota. Managing wetlands primarily for birds (or indeed other small groups of species) is problematic, as studies of other biota might reveal their different needs, which might run counter to bird-centered management and indeed endanger broader wetlands conservation efforts. For some time, wetland ecologists have acknowledged that there is an issue in the use of research focused on particular biota such as birds for general wetlands management. In 1999, Max Finlayson and D. Mitchell wrote that "monitoring has ... often been linked with autecological studies that have yielded a large amount of information on single or small groups of wetland plants ... or birds.... The value of much of this ecological research is not being doubted ... [but] difficulties arise in trying to effectively link ecological surveys conducted for one purpose to other issues. Changing this sectoral approach to research and monitoring is one of the key challenges when addressing steps to prevent further loss and degradation of wetlands."[112] For wetland ecologists, broadening studies beyond particular groups of species such as birds is important in assessing wider wetlands ecosystems and undertaking more effective wetland conservation.

Further, while biodiversity has been given increasing emphasis in wetlands research and management, this at times amounts to simply increasing the number of bird species surveyed.[113] For example, the biodiversity criterion for Fivebough and Tuckerbil Swamps is met though the number of bird species it is recorded as supporting, totaling eighty-three at Fivebough and sixty-nine at Tuckerbil.[114] Birds will continue to be important in environmental conservation and management in Australia partly for historical reasons, as there has simply been so much research on them, by both amateurs and professionals, that comparisons over time are better founded than for most other animals and plants.[115] There are also numerous social and economic rationales for foregrounding the importance of birds for particular wetlands, especially as potential tourism draws. For instance, Fivebough and Tuckerbil Swamps have become well-known places for bird-watching.

Over the last three decades, wetland ecologists have generated more information on wetlands animals such as frogs and plants such as river red gums to inform wetlands conservation geared toward their needs. For instance, in the Murray-Darling Basin, there has been growing research on the effects of altered water regimes on vegetation (especially floodplain species like river red gum), which clearly shows changes, as well as growing effects on invertebrates and native and introduced fish; however, "relatively little is known of the effects of altered flow regimes on other organisms."[116]

Just as those who undertook the stalled wetland survey learned, we need to be constantly attentive to not only what counts as a wetland but also whom and what wetlands are managed for. Drawing regularly on a greater variety of knowledge, rather than just documentary sources and research by scientists, could expand the variety of animals and plants as well as values that are considered by wetlands managers in dominant modes of conservation. The knowledges of Aboriginal people, farmers, and others have been increasingly incorporated into wetlands research and management, although much more remains to be done. This has included Aboriginal comanagement of protected areas, an early example of which is Kakadu National Park, established in 1979, which encompasses extensive Ramsar-listed wetlands. There burning regimes are part of the formal management process, a way of promoting habitat for a variety of plants and animals as well as meeting cultural obligations.[117] The Ngarrindjeri Regional Authority, which represents the Ngarrindjeri traditional owners, has shaped the management programs of Coorong National Park, which includes Ramsar-listed wetlands and is visited by birds subject to international agreements. This group has been directly involved in replanting and conservation of semi-aquatic plants, such as samphire, harvesting this and other plants for commercial sale.[118] Jessica Weir, David Crew, and Jeanette Crew have described the activities of the Yarkuwa Indigenous Knowledge Centre Aboriginal Corporation in forest wetlands management in southern NSW. This organization has endeavored to show that cultural practices fulfilling obligations to Country can help us rethink dominant approaches in environmental management that have often proclaimed nature to be wilderness and have led to the exclusion of Aboriginal people from Country, especially through the establishment of protected areas. These views of an externalized nature have conversely helped justify ways of living and farming that significantly contribute to river and wetlands degradation. At the same time, practices that take care of Country are important for the well-being of a variety of wetland species.[119]

Wetlands entered history in this period, gathering specific values and expertise, shaped by multiscalar politics and bird-centrism, and deeply connected to Australasian and Pacific circulations, both human and more-than-human. This history has had significant, but mixed, consequences for the way wetlands are understood and managed in conservation science and governments today.

SEVEN

RIPPLING

Capitalism, Seals, and Baselines

AT THE END OF THE NETWORKS OF RIVERS AND WETLANDS THAT compose the Murray-Darling Basin, salt and freshwater meet and mingle. Here, the Coorong lagoon and Lower Lakes—Lake Alexandrina and Lake Albert—form a liminal place. These shallow, dynamic water bodies, located in southeastern South Australia, mark the junction between the Murray-Darling Basin and the Southern Ocean, encompassing the mouth of the Murray River. Here, salt water and freshwater have been in a lively two-way relationship for thousands of years. This relationship has made life in this region possible. Scientists tell us that the area was open to the ocean until a peninsula formed sometime between six thousand and twenty thousand years ago. Since then this peninsula has sheltered the lagoon from the rough coastal waters of the Southern Ocean on the west. Until about two thousand to three thousand years ago the lagoon had two openings to the sea, in the north and south. At about this time the southern opening closed, leaving only the northern entrance (the river mouth).[1]

This is an area in constant flux, and the mouth of the Murray River has continued to shift location over the course of decades. When there has been enough rainfall, the freshwater of the Murray and Darling river systems has flowed into the ocean, carrying out to sea animals, plants, and microorganisms, some of which are washed back by waves onto the shorelines of the sandy coast. This dynamic promotes, for example, an abundance of pipis (similar to clams) along the beaches, nourished by the phytoplankton being washed ashore. At the same time, the sea has reached in, making some of the water brackish, a mixture of fresh and salt encouraging the presence of particular fish and other life forms. Indeed, some fish need to move between the brackish water and freshwater at different stages in their lives for breeding and development. This is a place of transition and change, where life flourishes in wide margins. The constant shifting movement between sea, freshwater, and land has fostered life, as well as livelihoods, in this region.

This mixture of fresh and salt has, however, undergone significant changes over the last 150 years, primarily though the construction of engineering works to support local as well as upriver agriculture and irrigation. From the late nineteenth into the twentieth century a series of drains were constructed just south of the Coorong lagoon for agricultural interests, in order to remove saline groundwater and surface water from this boggy area and improve farming prospects there. Prior to this, freshwater had flowed through surrounding land and creeks into the southern end of the lagoon, providing some freshwater inflow. The drains instead directed this water straight out to sea. Then, in the 1930s a series of barrages were built blocking the Lower Lakes from the Coorong and the mouth of the Murray River. These barrages were intended to protect upstream agricultural and irrigation interests by preventing the movement of salty water into the Murray River. Managers also used them to trap freshwater, carefully controlling its release from the river system and only fully opening them during the wettest years, when the river flow was high.[2]

The drains and barrages were built to control the mixture of fresh and salt water. The operation of these engineering works has created generally saltier water in the Coorong and Lake Albert, with consequences for fish, birds, other wildlife, and plants, as well as the livelihoods of people dependent on them.[3] This process has undermined Ngarrindjeri Aboriginal people's care for Ruwe/Ruwar, "lands, waters and all living things."[4] But in recent years the effects of changes earlier in the colonization of Australia that happened on islands out to sea are being felt anew in the Coorong, washing against the ripples from more recent capitalist interests such as irrigation as well as agriculture and fisheries. The sea again reaches in.

In 2007, a long-nosed fur seal (*Arctocephalus forsteri*) was seen in the Coorong lagoon. This species, sometimes also referred to as the New Zealand fur seal, is one of three pinniped species found in the wider area, the others being the Australian fur seal (*Arctocephalus pusillus*) and the Australian sea lion (*Neophoca cinerea*). All three species breed on islands, including Kangaroo Island, located off this part of the coast. However, before 2007 none of these seals was known to visit the lagoon, and long-nosed fur seals were only infrequent visitors to the adjacent Lower Lakes. Since that first sighting, however, long-nosed fur seals have become fixtures in the lakes and lagoon during their winter haul-out months, from about June to August (when they frequently leave the water to rest), with numbers at times reaching two hundred in the northern lagoon.[5] In fact, since the 1980s, the number of long-nosed fur seals in Australian waters has grown rapidly. Scientists have largely understood them to be rebounding after decimation by sealers in the late eighteenth and early nineteenth centuries and, crucially, reoccupying their former ranges.[6]

As their numbers have increased, however, the seals have become a controversial presence in the Coorong and Lower Lakes region, with differing views on whether they belong in these waters. Tour boat operators have benefited from the seals' presence, especially when they haul out on the warm concrete of the barrages. These charismatic animals are readily visible to tourists on the boats, who can watch them lopping into the water to catch fish. The fishways (also known as fish ladders) at the barrages, built to aid fish migration, have become known locally as "seal McDonalds" because seals take advantage of the fish moving through them for an easy meal.[7] However, it is the fish caught in the nets of commercial fishers that appear to be primarily attracting and keeping seals in the area.

For many local fishers and some Ngarrindjeri people, the seals are an unwelcome presence. These animals are seen as undermining the economic viability of the Coorong and Lower Lakes Fishery, by taking or maiming fish as well as damaging nets. They are disliked for their attacks on other wildlife in the lagoon. Both groups argue that these seals were never in the Coorong lagoon and do not belong there now. In the words of one fisher, "All of a sudden we had this top predator in the fisheries and in the area that was never there."[8] Some have argued for the seals to be culled. However, long-nosed fur seals are currently protected under state and federal legislation. While fishers and Ngarrindjeri people argue that these seals have no history in the Coorong, scientists continue to argue that the seals are

FIG 7.1 Goolwa Barrage is one of five barrages that divide the waters of Lake Alexandrina and Lake Albert from the Lower Murray and Coorong. Lounging and swimming long-nosed fur seals are just visible. Photo: Emily O'Gorman, 2014.

reoccupying their ranges from the days before seal hunting began, and protective legislation needs to be upheld.

Imagined ecologies—past, present, and future—are created and put to work through this dispute in the Coorong. A question of central importance in determining whether the seals belong is whether they were in the area prior to the sealing period. Appeals to historical baselines by managers and scientists have become a focus for resolving this controversy. This chapter examines the differing views and relationships with seals, focusing on the prominent role played by history and historical baselines, as well as some consequences of this reductive view of history. It argues that the seals' dynamic response to changing management regimes, as they are emboldened in the absence of human violence against them, undermines rationales based on presealing conditions and baselines. In such contexts, the past should be a guide but not a goal.[9] Ultimately, this chapter examines how historical practices and orientations have come to matter in this dispute.

In multiple and complex ways, the nature and extent of sealing in the past is shaping the current controversies over long-nosed fur seals in the Coorong lagoon. This chapter thus engages with historical sealing as a spatial and temporal rippling out, the effects of which continue to be felt in multiple places, as well as centuries later. A key aspect of past sealing that has shaped its ongoing effects is that it was an extracting capitalist industry. It helped create waves that have washed against contemporary capitalist industries like fishing, tourism, and upstream irrigation. Seals have inhabited all of these capitalist industries and their infrastructures, alongside their associated empire- and nation-building projects, in various and at times unexpected ways. They have been seen as a resource, as a pest, and as a spectacle, cocreating these more-than-human histories as "lively capital" or commodities and as threats to other sources of "lively capital" such as fish.[10] The particular behaviors of the seals have shaped each of these understandings, as their mobility and occupation of liminal zones have created both opportunities and profound difficulties for people.

This chapter particularly focuses on the contemporary experiences of fishers in the region. As seals are drawn to fishers' nets, they open up important questions about the ongoing consequences of past sealing, as well as questions of contemporary biocultural belonging. This situation brings historical legacies into conversation with larger national narratives of nativeness and invasiveness. In so doing, the chapter examines how broader changes on the Coorong and Lower Lakes have added pressures that further intensified experiences with the seals. It asks, Within this place of movement and change, what does it mean to belong, or conversely to be positioned as invasive? And what happens when once wide margins become narrow sites of control and capital accumulation? Importantly, the pressures created by the seals have encouraged some commercial fishers into new relationships with other species. Long derided as a destructive introduced species, freshwater European carp are being revaluated as a commodity by these fishers, as they seek alternative income streams due to pressures from drought and seals.

In many ways, the circumstances discussed in this chapter bring to the fore many hallmarks of the Anthropocene. Here, past and more recent efforts to support human mastery over nature continue to wash against each other, carrying all those involved into new, ethically and politically important, biocultural terrains. In this situation the past carries significant weight in management decisions, shaping notions of belonging that in turn powerfully influence life and death possibilities.

Engaging with these issues, this chapter develops a concept of rippling capitalism. In doing so, it aims to account for the temporal and spatial rippling out of sealing and other industries. In this regard the concept of rippling capitalism put forward here engages with and extends anthropologist Anna Tsing's notion of frontiers of capitalism and cultural geographer Rosemary-Claire Collard's work on disaster capitalism. Rippling capitalism is characterized by a particular temporal structure, in which past capitalist forms ripple forward and continue to be lived with in surprising and consequential ways. Crucially, tracing these ripples helps historicize more-than-human relationships as nuanced, dynamic, and complex. This is in contrast to the frequently ahistorical and idealized representations of these relationships in conceptualizations of historical baselines in environmental management.

SEALS AND CAPITALISM

As I walked down the driveway to meet Tracy Hill, I passed a number of fishing nets that had been strung up along the fence to dry. All the nets seemed to have large holes in them. I had already spoken to a few people in the area about the rise in the number of long-nosed fur seals in the Coorong and Lower Lakes and so wondered whether these were nets damaged by the seals. When I asked Tracy, she said, "If they've got big holes in them, it'll be the seals."[11] Tracy and her husband, Glen Hill, run a fishing and processing business focused on the Coorong and Lower Lakes. They have one of the larger fishing licenses in the area—composed of two licenses—which allows them to fish in multiple places at the same time if the fish numbers and conditions are right. Most of the fishing they do requires nets, and as Tracy noted, "if you've got great big gaping holes in your nets, then you're not going to catch fish." The nets are expensive, each costing one to two hundred Australian dollars to make. The seals have been attracted to the nets because of the fish they contain, and Tracy described seals not only biting the fish caught in the nets but grabbing and pulling at the nets in the water as they "rip the fish out of them." The seals sometimes do this while the fishers are trying to bring the nets into their boats. Tracy noted that "the harder and faster you pull the nets in by hand, the harder and faster they'll pull the fish out."[12]

FIG 7.2 Fishing nets damaged by seals. Photo: Garry Hera-Singh, 2020.

Tracy explained that she and Glen use gill nets to catch mullet in the Coorong, a fish prized for the sweet flavor it gets from the lagoon, with a relatively high market value. They can harvest up to ten boxes when there are large schools of fish. However, when mullet numbers are low, the seals often turn up. As Tracy explained, "When there's not a lot of fish around . . . you might only get a quarter of a box at a time, and then when the seals turn up, you've just got to pack up and come home." The seal presence also increases labor in watching over the nets: "You can't catch mullet when you've got thirty seals turning up on your nets at night . . . so you can't leave them [the nets] out." In the past, when mullet numbers were low, Tracy and Glen would concentrate on other fish, such as mulloway in the ocean or flounder in the estuarine areas. For all fish, however, they would normally have left nets in the water and regularly harvested the catch. But Tracy noted, "That's gone out the window now," because seals also damage these nets and fish catches, especially the nets placed in the estuary to catch the flounder. In short, the damage the seals cause to the nets, and their maiming of any fish caught, have undermined Tracy and Glen's ability to effectively work their fishing license.

174 | CHAPTER SEVEN

The Coorong and Lower Lakes are a multispecies fishery, in which fishers catch a variety of species, moving among the ocean, Lower Lakes, and Coorong. Tracy noted that "we've always survived on being able to shift effort from one species to another, either when they're most abundant and something else isn't or [when] the conditions weren't right. . . . So the advantage of multispecies fishing isn't there, because the seals don't allow you to fish for the other species when something isn't abundant." As a result, the seals have had a significant impact on their business. In a short period their business has reduced its turnover from approximately A$300,000–400,000 to less than A$200,000, with the most profitable time being when the seals leave the lagoon and lakes during the summer months. After expenses there is little left to live on and nothing to reinvest or improve the business.[13]

In this region, the impact of seals has come on the back of the effects of other hardships, in particular the Millennium Drought, which began in 2000 and lasted for a decade in the region. This drought had significant local economic effects, forcing thirty dairies to close because there was not enough water for their operations. During this entire period, the barrages remained closed, not letting any freshwater into the Coorong or the Murray mouth estuary. In parts of the Coorong the water became saltier than seawater, causing fish numbers to drastically decline and the water levels in the Lower Lakes to drop to such an extent that getting out to the water to fish became difficult.[14] The Coorong has remained hypersaline. As one fisher described it, it is so salty that it's "impossible to sink. If you fell overboard you wouldn't go to the bottom in parts of the South Lagoon in summer. . . . You'd actually float."[15]

While the Millennium Drought created significant and lasting hardship, many in the region assert that the barrages have had the biggest influence on the changing conditions in the Coorong and Lower Lakes. This is something fishers experience through reduced catches as well as observations of fish and other animal behavior. Both of Garry Hera-Singh's grandfathers were fishers in this area, as he is now. He told me that both of his grandfathers had said to him that "the single biggest change in this fishery from a fisheries perspective was the building of the barrages. Completely changed everything."[16] The barrages stopped the easy flow of fresh and salt water across wide margins. Only 10–11 percent of the estuarine area that was there in his grandfathers' early fishing days remains. This means that, for Garry, "production is 10–11 percent of what is was historically."[17] Some species of fish have declined more than others. Diadromous and catadromous fish that rely on moving between salt and freshwater systems to grow or reproduce have been most severely impacted. Garry explained:

> The Lakes system . . . was predominantly estuarine for most of the time, and then all of a sudden fish can't get back to their habitats. And yet for millennia these fish have been going backwards and forwards . . . laying their eggs or doing whatever they need to do to reproduce and then all of a sudden there's a barrier that says no. It's okay in wet years because the barrages are lifted and the water's allowed to escape, and the animals do have migratory access both ways. But because the focus over the last seventy-five years has been irrigation [in upstream states like NSW and Victoria], about not letting any water over the border into South Australia, the environment has suffered and all the animals that live in it.[18]

In recent years, fishways have been built to assist fish in getting over the barrages, which has helped some of these animals, but not all fish use the ladders, and connectivity through flows is intermittent.[19] The impacts of the barrages have been further compounded by reduced inflows in the southern end of the lagoon caused by the drainage works constructed during the late nineteenth and twentieth centuries.[20] Many people I spoke to in the Coorong and Lower Lakes area recalled seeing in their childhoods an abundance of birds and other animals, which have gradually declined. In this context the seals represent another significant pressure on lives and livelihoods in the region.

Like Tracy and Glen, other fishers have experienced financial impacts associated with the seals. These are more acute for less established fishers, who often have less capital or spare equipment. Garry noted that after he paid off his mortgage some years ago, he started buying nets and other equipment "rather than putting money in the bank." Since the seals have become a problem for fishers, he has given away six bales of net to help less established fishers who cannot afford to replace the nets damaged by the seals. He also told me that in 2016 "we had two fishers on high alert for suicide," primarily because of "the impact of the seals." Some fishers found themselves in a situation where they could not earn money through fishing because they had no usable nets, they could not afford the annual payments for their fishing licenses, and they were unable to gain a loan from a bank. In 2016, fishers in the region collectively estimated that seals were causing losses of approximately two million Australian dollars in saleable fish each year, over a quarter of the wharf value of the fishery's total catch in 2014–15.[21]

The financial implications have clearly created significant emotional stress. At the same time, fishers have found the behavior of the seals distressing in other ways. Many fishers have reported seeing seals attack waterbirds in a way that seems wanton and cruel. For example, they have seen seals biting the chests and bills of pelicans, as well as ripping the heads off ducks and other smaller waterbirds. Their maiming of fish seems wasteful to many fishers, as at times the seals appear to be interested not so much in eating the fish as in injuring them. Tracy described this behavior: "They don't just pull a fish out and eat it. They just go chomp, chomp, chomp, chomp, chomp, chomp [taking just one bite out of each fish]. Or they will just pull it out and drop it.... The [state environment] minister should be concerned about this just because of the sheer wastage of fish, let alone the dead birds and what impact it might be having on the ecosystem." Garry noted that "it's like a fox that gets into a henhouse and kills twenty chooks [chickens]—can only eat one but kills the whole friggin' lot. Well, seals are exactly the same."[22] Many fishers have taken photographs of injured wildlife and maimed fish, documenting the damage done by the seals to report to the state Environment and Fisheries Departments. But some have stopped doing this. Tracy explained to me that she had become tired of taking pictures: "It's distressing.... [Glen] comes home some nights, and it's like he's been to war. They'll pull the nets out of your hands."[23]

According to seal biologists, this sort of behavior by long-nosed fur seals is characteristic of juveniles (but not pups), and indeed many of the seals in the Coorong and Lower Lakes appear to be juveniles or subadults.[24] Seal biologist Tracey Rogers told me that young seals practice hunting through this sort of play behavior.[25] As the term "play" implies, the seals' objective is probably not to eat the birds or the fish, nor is it a given that they are primarily or intentionally honing their hunting skills. It is rather a game in which hunting tactics are incorporated and developed in a dynamic set of behaviors and learnings.

Seals are opportunistic feeders and happily move between salt and freshwater in search of food sources. Unlike the fish, they're able to easily get over the barrages. Young seals (both males and females) are, however, typically the ones who venture into new areas, testing out new sites to visit for winter haul outs. The seals' presence in the Lower Lakes, their exploration of adjacent water bodies such as the Coorong, and their experimentation with haul-out sites such as barrages, can be understood in terms of this demographic and the opportunistic nature of their movements, in which feeding and safety from predators are important concerns. The juvenile long-nosed fur seals that come into these waters are understood by scientists to be part of a breeding colony on Kangaroo Island, located off the South Australian coast, approximately one hundred kilometers from the mouth of the Murray River. By 2015, this and several nearby islands had become the main breeding sites for the seals in Australia.[26]

Under growing financial and emotional strain, in 2015 the fishers began to call for the state government to cull the seals. Although long-nosed fur seals are listed as being of "least concern" on the International Union for Conservation of Nature's Red List, this species, along with other pinnipeds, is protected under the South Australian National Parks and Wildlife Act 1972 and as a marine species under the Commonwealth Environmental Protection and Biodiversity Conservation Act 1999.[27] Under these laws, people cannot kill, harm, or approach the seals. These laws would not prevent the state government from authorizing a cull of the seals, but in this case the state government refused to do so. Its position was based on the common belief that because seal numbers were still recovering from sealing in the late eighteenth and early nineteenth centuries, seals should be understood to be reoccupying their former ranges. This has become something of a standard position for seal biologists since long-nosed fur seal numbers began significantly increasing in the 1980s. For instance, South Australian–based seal biologists Peter Shaughnessy, Simon Goldsworthy, and Alice Mackay noted in 2015 that "fur seal populations in southern Australia were heavily exploited by colonial sealers between 1801 and 1830, resulting in major reductions. Numbers remained low for 150 years, then slowly built up and new colonies established across their presumed former range."[28]

Instead of authorizing a cull, the state government provided some financial relief to the fishers by decreasing licence fees and increasing relief days. However, the local member of Parliament supported the fishers' call for a cull, arguing for an Overabundant Native Species Management Plan on the basis that long-nosed fur seal numbers in South Australia were over a hundred thousand, the presealing historical baseline estimated by biologists and historians.[29] More importantly, many local people believed the distribution of seals to have changed. Both fishers and Ngarrindjeri people argued that while there might have been more seals in the region prior to sealing, they were not generally found in the Coorong, and they were only occasional visitors to the Lower Lakes. As a result, they feel that the seals do not belong in these places now.

Some Ngarrindjeri people have also become deeply concerned by the behavior of seals, as they maim and kill animals that are important in kinship networks and responsibilities. Ngarrindjeri Elder Peter Mansfield-Cameron told me that "they're not a nice thing out here because they rip everything apart."[30] In 2015, Ngarrindjeri Elder Darrell Sumner killed three long-nosed fur seals in the area after seeing a number of pelicans—with which he has a close kinship relationship—being maimed or killed. His actions were in contravention of state and federal law. In a statement to a local newspaper, though, he noted that he would not "follow white man's laws on that."[31] In his view, "The seals have never had a history of being here."[32] In support of the argument that seals were not common in the Coorong and Lower Lakes prior to sealing, Elder Peter Mansfield-Cameron has pointed to the lack of archaeological evidence for seals in the area, noting that if seals were present they would have been killed by people, and traces would show up in this record. Further, he argued that seals have not been totems for his people. Seals do not appear in the many stories that describe the creation and spiritual importance of the landscape.[33] Indeed, they only appear in one story, in which Ngurunderi, the hunter—after chasing Pondi (Murray Cod), who created the Murray River as he traveled westward toward the sea—fought with a seal near present-day Victor Harbour. Ngurunderi smashed the seal's head with a rock, and the seal's eyes created a rocky outcrop.[34] Colonial records, albeit created when sealing had already been occurring off the coast of South Australia for a number of decades, do not refer to seals as common in the area, and no records have been found of sightings in the Coorong. One notable exception regards the observations of surveyor and explorer Charles Sturt, who wrote in his diary in 1830 that seals "frequent the lower parts" of Lake Alexandrina, though he makes no mention of them in the Coorong.[35]

While many locals continued to argue for a seal cull and to question the historical baselines used by scientists and managers, the state government has insisted that a cull would be ineffective and give rise to significant public relations issues. Some Ngarrindjeri Elders have recently presented another proposal: that the seals be relocated through a catch-and-release program.[36] Others have suggested sustainable harvests of the seals. However, the state government has not taken up these possibilities.

In some ways, the various capitalist industries involved here—sealing and irrigation, even fishing and tourism—might all be considered frontiers of capitalism. Tsing proposed this concept in the context of her research into the global flows of capital that supported mining in Indonesia in the 1990s, which she argued harked backward to, as well as driving forward, the type of frontier extractivism evident in European processes of colonization. She argues that "frontiers are not just edges, they are particular kinds of edges where the expansive nature of extraction comes into its own. . . . [T]hey are projects in making geographical and temporal experience."[37] Here Tsing is interested in the continuation of a colonial extractivist narrative and process through various capitalist industries. Similarly, in her discussion of disaster capitalism, Collard has sought to understand the effects of the 1989 oil spill off the Alaskan coast on endangered sea otters in terms of longer histories of capitalism in the region. She examined the sea otters' "orientation in relation to capitalism and the state," beginning with their mass killing for the fur trade in the eighteenth and nineteenth centuries and tracing their declining numbers through successive capitalist industries.[38] This is a different story from that of the long-nosed fur seal in the Coorong, whose numbers have not continued to decline like the sea otters but seem to be recovering, with mixed consequences.

The seals in the Coorong bring us into a different engagement with capitalism and temporality from those put forward by Tsing and Collard, drawing attention to a dynamic rippling out of capitalism in the Anthropocene. The seals ask us to consider not just similar processes of extractivism that supplant each other in temporal succession but the ongoing effects of past capitalist industries such as sealing, which continue to produce transitions and changes in the lives and behaviors of seals today—and consequently other lives as well, including those of humans—in important and unexpected ways. While seals (like otters) inhabit different capitalist processes in different ways, shaping their lives and deaths, these capitalist forms do not transition neatly and definitively into one another, one after the other. Rather, they ripple and fold into one another. In an important sense, sealing is not yet over.

The concept of rippling capitalism put forward here is novel in its attention to the ongoing legacies of particular capitalist industries as they continue to influence the geographical and temporal experiences of humans and nonhumans. Attention to these ripples is important in historicizing more-than-human relationships as dynamic and shifting rather than ahistorical, as is frequently the case in the conceptualization and use of baselines in environmental management. My focus here is on contemporary concerns about seals in the Coorong, in which the waves of change brought about by rapacious sealing in the late eighteenth and early nineteenth centuries are still being felt by seals, humans, and many others. These have washed against other forms of capitalism that also significantly shape this region, such as irrigation on the Murray and Darling Rivers and the fishery. The way these effects have played out has been shaped by seals themselves as well as conservation imperatives, Indigenous knowledges, and these contemporary industries. Indeed, even the extreme uncertainty that surrounds the numbers and exact locations of past seal populations is partly a product of the patchy records generated by the particular commercial form of sealing that took place, which was fast-paced, violent, and at times secretive.[39] Sealing practices have created diverse absences and presences that continue to be felt in unpredictable and unexpected ways.

ABSENCES, PRESENCES, AND BASELINES

The mass killing of seals and sea lions, primarily for their pelts but also for their oils, happened relatively rapidly off Australian coasts. Beginning in Bass Strait in the 1790s, sealing was all but over in southern Australia by the 1830s, with any surviving seals located in places so inaccessible to people that it was often no longer profitable to find and kill them. Most sealing took place in the thirty years starting in 1800, in a capitalist frenzy that boomed so quickly it became Australia's first export industry.[40] In South Australia, the center of the sealing industry was Kangaroo Island. In the early 1800s a group of male European sealers, Aboriginal women, and others established a community on the island, where there were large breeding colonies of fur seals, as well as a good supply of salt with which to preserve the pelts. This latter resource meant that Kangaroo Island remained an important stop for sealing boats in southern Australia even as seal numbers on the island dwindled.[41]

The cultures of sealing that quickly developed in this period were violent and created significant upheavals for many Aboriginal people in Tasmania and mainland Australia. Aboriginal women, including Ngarrindjeri women, were kidnapped by sealers and taken to seal colonies located on islands, where they were expected to assist in sealing and provisioning food. These kidnappings and the abuse of Aboriginal women were part of and at

FIG 7.3 Kangaroo Island was a center of sealing in Australia in the late eighteenth and early nineteenth centuries. Image: William Woolnoth, *View on the North Side of Kangaroo Island*, 12 February 1814, Image 30328102131751/3, State Library of Victoria.

times instigated further frontier violence, for instance as Aboriginal women's families retaliated against European sealers and colonists. Yet as Lynnette Russell, Lyndall Ryan, and other historians have argued, Aboriginal women also inhabited and shaped these situations with complex agencies. The interactions and domestic spheres that resulted were molded by gender relations—not just racial ideologies—in which Aboriginal women exerted considerable influence. The island locations of sealing enterprises further shaped these relations, as the communities that developed were often isolated for long periods and needed to be somewhat self-reliant. Indeed, captains of sealing boats intentionally stranded European men on islands to undertake sealing and did not return to collect them for many years.[42]

Sealing was a fast-paced, exploitative industry that generated sparse historical records. In recent decades biologists and conservationists have sought to understand the impacts of sealing on seal populations to assist and evaluate their recovery. They have looked to historical records to gain information on the population sizes and locations of the seals prior to sealing. One key study was undertaken by John Ling in 1999, in which he analyzed a variety of historical sources generated by sealing, including sealers' and others' accounts, log books, and cargo inspections. Ling estimated that since sealing began, "almost 1.4 million fur seal skins were harvested around Australia's southern coast, New Zealand and at the adjacent subantarctic islands," with most taken before the 1830s.[43]

Trying to reconstruct seal numbers for the purposes of contemporary conservation from sources generated for very different reasons has posed numerous problems. Sealing records were at times deliberately vague and patchy, as sealers dodging the law as they undertook illegal activities did not want to give away their locations. Sealing boats often visited several sites to drop off teams and collect pelts, but their cargo was only inspected once they came to port, and the inspectors only noted the total contents. Inspectors' reports gave vague estimates of the numbers of skins on board boats, using imprecise terms such as "a bundle of skins," and their estimates further failed to account for skins that were lost or spoiled before loading.[44] In addition, some boats did not come to port before exporting in and beyond the British Empire, with some cargoes shipped direct to overseas markets, including parts of Asia. The skins of different kinds of seals were often not differentiated for the purposes of cargo inspection or ship logs. The fervor and lack of regulation in the sealing business created not only an absence of seals but gaping holes in historical records as to how many and which seals, or rather seal skins, were harvested from where.[45] As a result, any baselines of seal ranges and numbers prior to sealing are far from certain. In some places, like Macquarie Island, the killing was so complete that virtually no seals were left, and there are no meaningful records to identify the seals that were there. While seals have reinhabited the island, there is no way of telling whether these are the same kinds of seals that were there when the first sealers landed.[46]

Researchers like Ling have made conservative estimates of populations based on the records made that have survived. Ling tallied the fur seal skins known to have been shipped from Kangaroo Island between the commencement of sealing there in 1803 and the last shipment in 1834, producing a total of 98,973 skins. He has argued that the dominant species on the island was probably long-nosed fur seals.[47] He put the total number of fur seal skins shipped from South Australia up to 1844 at 99,023, which included the

Kangaroo Island seal skins.[48] These totals are conservative, given the patchy nature of the records, and do not differentiate between Australian and long-nosed fur seals. Yet in the absence of better records they have come to stand in for historical baselines of total fur seals in South Australia.

Seal ranges prior to sealing are also difficult to determine from these records. Ling argued that "in the absence of any definite information and [given] the fact that there is little likelihood of any being forthcoming, it must be tentatively concluded that the same species occupy the same ranges today as they did when discovered by early navigators and sealers more than two centuries ago."[49] Yet, as the situation in the Lower Lakes and Coorong indicates, when this assumption is contradicted by other understandings of the past, and the belonging of the seals is questioned, such reliance on historical assumptions becomes difficult to defend.

Fur seal numbers remained low into the twentieth century. They received some legal protection in South Australia in 1919 but only in some areas, and they were often killed by fishers who intermittently urged for them to be systematically culled.[50] In South Australia, fur seals were fully protected under state legislation with the introduction of the Fauna Conservation Act 1964 and later under the National Parks and Wildlife Act 1972.[51] These acts were part of a series of conservation bills introduced across Australian states in the 1960s and 1970s that aimed to redress the losses of fauna and flora since British colonization, including the mass killing of animals like seals and sea lions. Biologists and conservationists hoped that these acts would help fur seal numbers recover. These legal protections around seals were directly or indirectly shaped by sealing and more widely by capitalist and colonial cultures of mass killing of animals. They reflected changing values associated with native animals in the ethos of the environmental movement and led to greater government regulation of killing. In the 1980s, both Australian and long-nosed fur seal numbers began to increase. The reasons for this relatively sudden growth, beyond their ongoing protection, are unclear.[52] In the 1990s, long-nosed fur seals were actively colonizing Kangaroo Island, and in 1994 biologists estimated the total Australian population to be 34,700.[53] By 2015, this estimate was closer to 100,000 in South Australia alone.[54]

From at least 2007, long-nosed fur seals started frequenting the Lower Lakes, hauling out on the barrages and eventually visiting the Coorong, likely only for a few days at a time.[55] Fur seals can travel quickly, covering distances of five hundred kilometers in a day. Traveling to this estuarine area from breeding sites on Kangaroo Island, located about one hundred to three hundred kilometers away, is a fairly easy trip. Young fur seals are adventurous and characteristically explore new places, and it seems that these were the ones to test out the possibilities of this area. After they are about a year old, pups have often been taken by their mothers to feed off the continental shelf and in the waters of the sub-Antarctic. Most stay with their mother for about another year. At roughly four years of age, female seals start breeding and often move between the island breeding colony and the sub-Antarctic waters for a ready supply of food. All fur seals are opportunistic feeders, in the language of biologists, but it is mostly juveniles who venture into new areas with good supplies of food, especially fish of almost any kind.[56] The Lower Lakes and Coorong therefore presented an ideal place to visit during their winter haul out, with a ready supply of fish in fishers' nets, thrown off boats, and at the barrages.

The Lower Lakes and Coorong may be new areas for generations of long-nosed fur seals after the sealing era. With the changes that have taken place in this area over the last two hundred years or so, these places present entirely different prospects for these seals, who are experiencing a rapid population increase. The food sources presented by the fishers and barrages constitute one such change. Yet food sources would have been plentiful enough to tempt the seals prior to sealing and before this infrastructure was built. Another significant, relatively recent change, however, involves the legal protections that have largely prevented seals from being killed in recent decades. There is a possibility that these legal protections are changing seal behavior during a time of population growth, as the seals are now able to venture into coastal and estuarine areas in ways that they would not have been able to previously. Biologists speculate that seals have formed large colonies on islands because of their relative safety from land-based threats, primarily humans.[57] Prior to British colonization, it is possible that seals never became common in the Coorong and Lower Lakes because Ngarrindjeri people killed any adventurous juveniles or subadults they found testing out the nearby coastal water. Certainly it would have been easy to kill any seals who made it through the Murray mouth and rested on land. Indeed, prior to and at the time of colonization, coastal Aboriginal people killed seals in many parts of Australia.[58] Later, during early British colonization, sealers killed them en masse. When sealing slowed, various people, including fishers, would likely have continued to opportunistically kill any visiting seals. With more effective protection over the last several decades, this has changed.

Nevertheless, it seems likely that prior to sealing at least some seals traveled into the mouth of the Murray River, through Lake Alexandrina, and into the freshwater Murray River system, as they did after the sealing era. There is some speculation that seals who ventured into these river systems were the origin of the bunyip figure in some Aboriginal cultures. Over the last two centuries, there have been occasional reports of fur seals and leopard seals in the upper parts of the Murray River and even in the Darling River system.[59] Yet these sightings have been rare, and it seems that only the occasional individual ventured into these waterways, despite seals' ability to live in fresh as well as salt water.

According to this hypothesis, changes in human behavior—from killing to not killing seals—is changing seal behavior and distribution in unexpected ways. Indeed, fishers in the Coorong have commented that the fur seals behave brazenly, as though they are aware the fishers cannot harm them. The fur seals pull on the nets as fishers try to bring them into their boats, and occasionally they climb aboard the boats, with fishers noting that "they know we can't do anything."[60] For recent generations of long-nosed fur seals, then, the Lower Lakes and Coorong may present a place of safety and food, free from land threats. Further, it is free from ocean predators like orcas and sharks.[61] At the same time, competition with ocean fisheries and reductions in saltwater fish may also be encouraging seals to seek out other places to find food, a situation that may be exacerbated by the effects of climate change on fish numbers and distribution.[62]

In this dynamic context, a focus on static baselines becomes particularly problematic. Most fundamentally, it raises the question as to why past seal numbers and ranges should guide management goals so significantly. The exact numbers of seals in Australian waters at the time sealing commenced will never be known, and yet figures extrapolated from sparse evidence have become central to the seal controversy on the Coorong. This situation resonates with an argument by historians Peter Alagona and John Sandlos and biologist Yolanda Wiersma in the context of the United States, that "conservationists and restorationists too often adopt historical baseline data uncritically, using early accounts of wildlife numbers or ecological conditions without closely examining the manner in which these accounts were recorded, transmitted, and interpreted by past generations."[63] As in the case of seals in Australian waters, historical baselines often rest on fragmentary evidence, with species numbers arrived at through sparse records. There are also contradictory historical accounts that are locally specific, including from fishers and Ngarrindjeri people.

But no matter how accurately they are known, these pasts can never be re-created in the present and the future. This is an impossible task within dynamic relationships and wider ecologies. Indeed, the dynamic behavior of the seals themselves as they respond to changing conditions undermines the use of baselines. As such, there is no reason to simply assume that the number and distribution of seals found in this region in the past can still be supported in the present, in the context of radically transformed estuarine and coastal environments. This does not mean history is irrelevant or should be ignored in environmental management. Rather we need more nuanced and complex histories that account for dynamic and shifting behaviors of, and relationships between, humans and nonhumans, to inform environmental management. In the Coorong and Lower Lakes area, as elsewhere, it is important to recognize that these histories are key to understanding established and emerging relationships that include humans and nonhumans. They cannot be escaped and carry with them legacies that we all must grapple with. While history is important in conservation planning, as noted by Alagona, Sandlos, and Wiersma, its use within a narrow baseline framework that sets the past as a future target is problematic, as ecologies and societies are continually changing. The past should be a guide but not a goal. In all of these ways, the exploitative enterprise of sealing continues to ripple out through a nuanced array of absences and presences that shape complex questions over the biocultural belonging of seals in the Coorong and Lower Lakes area.

LIVING AND BELONGING IN THE MARGINS

When I visited the Coorong and Lower Lakes area in 2016 and 2017, the language scientists and managers used to describe the seals was changing. For a number of decades, perhaps the most common name for the species *Arctocephalus forsteri* in Australia had been New Zealand fur seals. Scientists and managers were now starting to use the name long-nosed fur seals in an attempt to dispel any notion that these seals had been introduced to Australia from New Zealand. Many people in the area, both fishers and others, felt strongly that the seals did not belong, not only in the Coorong lagoon but ultimately in Australia. Some used the name New Zealand fur seal to assert that this was their proper geographical range and to push for the state government to undertake a cull. For many years scientists had seen the name as misleading because these seals have long inhabited Australian waters.[64] Thus belonging is never simply a question of biology or culture in isolation but a terrain of contested biocultural meaning, with life-and-death stakes. Questions of belonging always involve biocultural belonging.

Sealers called these long-nosed, or New Zealand, fur seals "black fur seals," whereas Australian fur seals were called "brown fur seals," likely reflecting the sealers' primary interest in the qualities of the pelts.[65] By the early twentieth century, a number of common names were used for *Arctocephalus forsteri*, including Australasian fur seals, South Australian fur seals, and New Zealand fur seals. The name New Zealand fur seal appears to have gained increasing prominence in the mid-twentieth century, reflecting what had become the geographical concentration of these seals at that time. Its common usage was reinforced by scientific research on the differences and connections between geographically disparate, and at times now fairly small and isolated, fur seal populations and their possible categorizations into distinct species. By the late 1960s, the separate species classification of the sea lion (*Neophoca cinerea*) had been widely accepted. However, the identity of three possible species of *Arctocephalus* was still being debated. Three species had been described in the 1920s in publications by different people: *A. tasmanicus*, known as the Tasmanian fur seal; *A. doriferus*, known as the Australian fur seal; and *A. forsteri*, known primarily as the New Zealand fur seal. In 1968, zoologist Judith King published the results of her examination of pinniped skulls in Australia, concluding that the two main species of fur seals in Australia were the Australian and New Zealand fur seals, and the Tasmanian fur seal was in fact synonymous with the Australian fur seal. This article authoritatively established some important species differentiations as well as the vernacular names.[66]

By 2015, the popular name New Zealand fur seal had become such a feature in arguments for a cull that scientists proposed a new name. That year, Peter Shaughnessy and Simon Goldsworthy published an article in *Marine Mammal Science* in which they argued for long-nosed fur seal as "a new vernacular" for *A. forsteri* in Australia to "improve public perception of the species."[67] They explained that the two species of fur seals that bred off the coast of Australia had "[in] recent decades . . . been recovering from overharvesting in the 19th century," leading to "complaints in the media and to government departments from commercial and recreational fishermen about burgeoning seal populations." They wrote that support for culls of New Zealand fur seals in South Australia were "based partly on the uninformed belief that the local fur seal is feral and has been introduced from New Zealand" and that "the logic for that belief is that if the Australian fur seal is a local species in southern Australia, then the New Zealand fur seal must have been introduced."[68] However, they had "found no evidence that the New Zealand fur seal had been introduced to Australia and deduce that it was in the country long before Europeans began harvesting them in the late 18th century."[69] In fact, these seals had arrived in Australian waters sometime in the last million years.[70] After considering various options for a new common name for the fur seals, including Aboriginal names, of which they concluded there were many regional differences, they settled on long-nosed fur seal "because the extended nasal region ... sets it apart from the Australian fur seal."[71] This name perhaps also reflected the biologists' interest in the physiology of the seals.

In 2017, the name had already gained some traction but was not well accepted by fishers in the Coorong, who continued to question whether these seals really belonged in Australia, even if they were present off the coasts prior to British colonization. One fisher, Tracy, stated, "It's been sort of accepted that the seals don't belong in the Coorong and the Lakes. They've also said that the population of the New Zealand fur seal—because that's what they are even though they call them long-nosed fur seals—is probably back to ... [presealing numbers], so one hundred thousand plus, and that's one of the reasons they've come looking for alternative habitat is because where they were is used up."[72]

Continued use of the name New Zealand fur seal functions as an argument by fishers that these seals should be culled. Likewise, the scientists' push to change the name testifies to the recognized power of labeling a species as invasive in Australia and this country's long and problematic relationship with belonging. Species that arrived after British colonization in 1788 are often labeled as invasive in Australia and understood as detrimental to native species. Livestock and other commercially valuable species generally escape this label, despite their significant environmental impacts. Here, notions of nativeness have underpinned claims to a species' belonging, whereas invasiveness has indicated that a species does not belong. This division has a long history of informing systematic programs of killing that target species that don't belong. There is a different baseline concept at work here, one in which 1788 is used as a broad management goal.[73] Here too, though, the past may be better as a guide.

In recent years this approach to management and the conception of belonging in Australia have been questioned by biologists, social scientists, and humanities scholars, who argue that we need more complex understandings of particular organisms and of belonging, which take account of both the dynamic nature of ecosystems and the specific characteristics of newly arrived species.[74] Fishers too have been rethinking belonging but in economic terms. Here, belonging is framed as an economically useful term. We can see this in some of the fishers' underlying concerns that the seals threaten their livelihoods, suggesting that their lack of belonging is due to their economic disruption. There is a moral dimension perhaps here too, in that the seals are seen as cruel because they maim birds and waste food.[75]

While the belonging of seals is being questioned, the difficult economic position faced by some fishers—in part as a result of these seals—has led them to rethink the usefulness and place of a much-maligned invasive species in Australia: European carp (*Cyprinus carpio*). In search of a different source of income, Tracy and Glen have begun to catch and process carp, which are present in large numbers in the Lower Lakes. European carp, which are native to China, from there moving throughout Asia and Europe, entered the waterways of the Murray-Darling Basin starting in the 1960s and significantly extended their range during floods in the 1970s. The most prolific group of carp was introduced to these rivers from an ornamental fish farm in Victoria in the early 1960s, and they are now in most of the waterways in the basin. They are maligned by environmental and fish management as invasive pests that muddy the water as they feed along the riverbed and significantly reduce native fish like Murray River cod. They are also difficult to contain, as they can propel themselves over impoundments.[76] During the Millennium Drought, with little freshwater inflow coming down the Murray River, fish numbers in the Lower Lakes significantly declined, and carp were one of the few fish that remained. It was during this period that Tracy and Glen began to take a real interest in them. With water levels low in Lake Albert, the carp accumulated in one place, and the state government contracted Tracy and Glen to remove them. Tracy and Glen employed a carp exporter from Victoria, who taught them how to haul, harvest roe, fillet, and prepare carp "so you didn't have that muddy taste."

In addition to being disdained as invasive pests, carp are renowned in Australia as bad to eat, "bottom feeders" who taste of the mud in which they forage for food.[77]

Along with harvesting the carp in accordance with their government contract, Tracy and Glen continued catching mullet in the Coorong. Tracy had intended the carp processing to be a one-off: "I didn't think much more about it at the time because we were just catching mullet, plenty of fish around, and then the seals got really, really bad, and the mullet shut down." As freshwater started coming back down the Murray River, Tracy explained that there was a "transition [period].... It took about twelve months for the Coorong to sort of reset itself and the food chain to reestablish." This, in combination with the impact of the seals, meant that they needed to find another source of income: "We were starting to get a bit desperate. We were applying for jobs, you know? We were looking at going up to the Northern Territory for work." Prompted by a friend, Tracy started to reimagine the business in terms of the skills she and Glen had learned during the drought—that is, catching and processing carp. Tracy noted that "we were already heading down the carp path because we were looking to harvest large quantities for Charlie Carp fertilizer," but harvesting and processing carp for sale as food has now become a regular part of their business. They sell carp products at local markets and other outlets, and are now looking at international export. A major obstacle has been making carp products that people will eat, largely because of the stigma of carp in Australia. Tracy recalled thinking that "I know it's going to take a lot of educating and changing people's minds, but I know if we can get people to taste it, they will go 'wow.'"[78] Their continued interest in carp and their reevaluation of how this species might have an economic place can be viewed as another rippling out of sealing, which continues to shape lives and livelihoods.

Tracy began to experiment, making Thai fish cakes, sausages, and mince from carp flesh. Attuned to the marketing issues, she steered away from putting carp in the name of these products, instead promoting fish products low in calories and high in omega-3. This sort of making do has left Tracy and Glen somewhat reliant on carp at a time when environmental managers are seeking to reduce carp numbers and ultimately eradicate them from the river systems. Indeed, in 2016, politicians and some environmental managers proposed releasing the koi herpes virus among carp populations in these rivers to eliminate them. For Tracy and Glen, having just learned to incorporate the carp into their business, this was yet another blow: "And now they want to put a virus in and kill them all. And I'm going, 'I just started getting something out of them . . .' Getting the carp biomass down is important, but I've had investors ring us up and say they would like to invest in carp processing and carp exporting." The proposed release of the herpes virus has been controversial, with questions over its effectiveness and the consequences of a sudden death of large numbers of carp on riverine ecologies. For Tracy and Glen, going into the carp business was a matter of making do in bad circumstances. In such an environment, with multiple pressures, who belongs is a shifting and multiscale biocultural question. While in Tracy's view neither seals or carp belong in this area, it is clear that carp have started to gain an economic place among fishers, and yet for the state Department of Environment and Water, seals belong and carp have no place. Tracy noted that the carp processor they had employed during the drought was looking to move his business to the Coorong and Lower Lakes, but he could not get any government assistance, in Tracy's view because the government didn't want to support a carp business: "But there's jobs there, there's carp harvesting, there's carp processing, there's a protein source just going to waste."[79] Both seals and pelicans are also making do with carp, as both have incorporated these fish into their diets.

In the Coorong and Lower Lakes, a range of humans and nonhumans are finding ways to craft "life in capitalist ruins."[80] These are modes of getting by in difficult circumstances, significantly shaped by the ongoing ripples of sealing along with broader changes in the Coorong and Lower Lakes, including reductions in freshwater inflows and other transformations brought about by the barrages. Many of these modes of getting by may point toward alternative futures. As the margins have become narrower in the Coorong and Lower Lakes, people as well as nonhumans have developed innovative solutions.

Fishers have needed to find new ways to fish in order to address the problems presented by seals. Part of this has involved increasing their labor in watching over the fish nets and pulling them in when seals arrive. Fishers have also investigated using different nets, looking at adaptations made by fisheries in the Northern Hemisphere in response to seals. However, for a variety of reasons fishers in the Coorong have found that these options do not work in their environment. The state government has supplemented the cost of firecrackers used to scare seals away. While this has worked to some extent, the seals seem to have learned that the firecrackers do not physically harm them and have started to ignore them. Fishers are willing to adapt their fishing methods, but finding ways that are effective in the long term and financially feasible is proving difficult.[81]

Another species included in the multispecies fisheries are pipis, which live on the ocean side, nourished by the nutrients that flow out of the river mouth and are washed back onto shore by the waves of the Southern Ocean. Pipis may ultimately offer some financial relief to at least some fishers. Historically pipis have had a relatively low market value, often being used for bait or as a less desirable, cheaper source of food. However, recently they have become more valued as a sustainable food source. Garry Hera-Singh noted that "the catch [of the fishery] has been steadily declining since 2007, but the value is on the increase, attributable to an increase in value of the pipis for human consumption."[82] This meant that in 2016 the price rose from about 50 cents a kilogram to $10. Yet not all fishing licenses include pipis, and not all fishers are equipped to harvest them. There are also quotas on the number of pipis that can be harvested.

There are no long-term solutions emerging for the fishers, nor is there likely to be a singular answer. Indeed, if there were to be, we might worry that like the fate of seals in the past, this pressure might lead to unsustainable levels of harvesting with all of the attendant problems. Instead, fishers have increasingly needed to cultivate diverse and shifting practices for what Anna Tsing has called "gardening in the ruins," making do under multiple pressures.[83] The seals are significant among these pressures, as the ripples of sealing continue to move out in time and space, shaping human and nonhuman futures in a region that has increasingly narrow margins of overlapping fresh and salt waters. But the seals are tangled up with other pressures, from changing conservation and environmentalist agendas to the impacts of irrigation and drought. Getting on within the complex folds of a temporally and spatially dynamic Anthropocene is a challenging task, one that requires ongoing revision and experimentation. More-than-human histories can play a vital role in this work—but not by producing a notion of the past that might serve as a goal but is ultimately ahistorical as an unchanging point in time. Rather, we must articulate ambivalent guides through dynamic histories both to what has been and what might yet still be possible, opening our attention to the diverse agencies and relationships that have always coconstituted these watery places.

AFTERWORD

AS I FINISH WRITING THIS BOOK AT THE BEGINNING OF 2020, Australia is in the midst of another dry summer. This one has been particularly difficult, coming after a number of dry years with increasing competition over water and large bushfires in many places across the continent. In the Darling river system, rivers and wetlands are drying out, and it is unclear how all those reliant on them—from towns to birds and wetland forests—will make it through many more months without significant rain. This summer and last, thousands of fish, including iconic native Murray cod, died in a series of "fish kills." Water levels in the lower Darling River became so low and the weather so warm that algal blooms flourished, then cold fronts hit, killing the algae, and then bacteria eating that algae removed oxygen from the water, making the environment toxic to the fish. Last summer, some local farmers tried to explain the dire conditions in the national and international press by emphasizing that many of the Murray cod were fully mature. Able to live for approximately fifty years, the fish had experienced many water levels and changes, but their world had become a place where they could no longer survive.[1] The bodies of the dead fish told a more-than-human history and possible future.

Heralding yet another dry summer, and what many saw as a sign of fires to come, in October 2019 a lightning strike started a fire in the dry Macquarie Marshes. These fires burned more than three thousand acres, 90 percent of the reed beds. There is little water in upstream Burrendong Dam to release to help the marshes recuperate.[2] If rain and even floods come soon, the marshes might recover, slowly. While this is undoubtedly the best-case scenario, government water reform and action on climate change might then lose momentum as constituents' attention turns to other matters, as has happened before. That is until the next drought, which might be much worse. Anthropogenic climate change is one of the reasons that we need longer-term changes in government approaches to water and wetlands management.

Anthropogenic climate change is making droughts more severe and crucially exacerbating the ongoing effects of narrow water management in the Murray-Darling Basin. Over a decade ago, Australian government researchers predicted a warming of the climate in inland Australia of 1–1.2 degrees Celsius by 2030 and 1.8–3.4 degrees Celsius by 2070, driven by anthropogenic climate change.[3] A separate study predicted that these climatic changes would interact with weather systems to alter surface water availability in the Murray-Darling Basin, which would either increase by up to 7 percent or decline by up to 37 percent.[4] A recent report indicated that freshwater inflows to the Murray River system over the last two decades have been half of what they were during the previous century, signaling that large declines in water availability may be likely in the years to come.[5]

While the net effects of climate change on freshwater availability remain uncertain, droughts are already and will likely continue to become longer and drier, and floods fewer but more severe. This means that wetlands and river systems will experience longer dry periods punctuated with more extensive periods of inundation. For wetland animals and plants adapted to more nuanced water variability, this will create big and cascading changes. For example, under such conditions river red gum trees will become stressed, because they rely on intermittent flooding and drying and do not respond well to long periods of either drying or flooding. Some animals prefer living in the live trees, while others prefer inhabiting the dead trees, and conditions will favor the latter. Further, during long dry periods, many wetlands will become more like terrestrial systems, favoring that biota. Importantly, they will not act as drought refuges in the way they have in the past.[6]

The effects of climate change on wetlands around the world will vary greatly, with, for example, storm surges either damaging or being buffered by coastal wetlands in various places, or creating new wetlands that may or may not be welcome. In general terms, as with inland Australia, a reduction in rainfall in already dry areas will have a major impact on wetland ecologies, especially as refuges for many species. For example, greater drying of arid wetlands in the United States will reduce connectivity for migratory species reliant on them, such as a variety of birds.[7]

In Australia, Aboriginal groups throughout the basin have been vocal in demanding changes in management of water and wetlands and have insisted that their voices be heard as their Country suffers. Muruwari and Budjiti man Bruce Shillingsworth spoke out on a national current affairs television program about the mounting issues in the Murray-Darling Basin: "The impact of the water mismanagement, and the corruption, and the corporate greed, and capitalism in this country has killed our rivers. They've killed our communities. . . . Where are our rights to water?"[8] Indeed, the recent Millennium Drought revealed that Aboriginal people and communities were disproportionately affected by water shortages, as access to safe and affordable drinking water became more difficult. Further, important cultural sites and species were put at risk.[9] The effects of climate change combined with narrow water management will further exacerbate these effects on Aboriginal communities and Country.[10]

The scope for reimagining wetlands and rivers in the basin is being curtailed by climate change as the possibilities become narrower. Yet imaginative responses are needed now more than ever. Indeed, wetlands offer pathways toward action on climate change. Serving as carbon sinks, drought refuges, and sites of cultural connection, wetlands can provide important means for addressing some of the many socioecological aspects of climate change. The way people address and adapt to these changes will be crucial, including making room for new wetlands that might form and finding new ways to live with and support the remaining ones. Continued narrow management of water in the Murray-Darling Basin, primarily for irrigation and other farming needs, will further reduce flows to wetlands and prevent possible new wetlands, albeit with quite different water regimes, from forming.

Although climate change presents a new challenge, important legacies are evident here, and many issues are ongoing. Crucially, these need to be understood as multicultural and multispecies histories and futures, shaped by dynamic, diverse agencies. This approach widens our view not only of the past but also of possible futures. As this book has shown, approaches that have pitted the "nature" of wetlands against the "culture" of agriculture have failed to recognize their entangled more-than-human agencies and have prevented more just and sustainable forms of environmental management.

Importantly, the voices of Aboriginal people need to be listened to and indeed occupy a prominent space in environmental management, not only to address past injustices but also to ensure healthier Country and communities into the future. Further, an engagement with the long histories of Aboriginal people's connections with and uses of wetlands overturns narrow views of these places as damaged by human practices. Instead, we need to acknowledge the vast array of potential interactions people can have with wetlands.

This book has sought to show that far from being static or pristine nature, wetlands are places of multifaceted human and nonhuman encounter. Wetlands are sites in which issues of race, class, and gender have taken, and continue to take, shape within dynamic human and nonhuman relationships. How wetlands have been used and valued, and the contention over them, have been deeply embedded in gender and socioeconomic relations, revealing the ongoing pressures and changes facing many groups and individuals. Managing wetlands as cultural spaces, with lively human and more-than-human histories and futures, would look very different from most current models. Indeed, doing so could lead to more socially just and sustainable models. Further, various wetland animals and plants have responded differently to pressures. Their dynamic lives and behaviors mean they have often moved across and between wetland, agricultural, urban, and wider landscapes in ways that need to be better accounted for by managers, farmers, and others. We need to keep revising the meanings and values we associate with watery places and the category of wetlands, as these continue to inform their management. Importantly, as we work toward better futures for wetlands, in ways that support their rich, diverse human and nonhuman worlds, the past should be a guide but not a goal.

Many of the issues raised in this book, such as those related to the legacies of colonization, the management of invasive species, and baselines, are relevant to places beyond Australia. The book has grappled with economic, social, and technological changes, as well as climate change. Indeed, Australian perspectives on uncertain water and climate regimes are important contributions to global discussions as we enter an era shaped by changing climate regimes and the responses of dynamic human and nonhuman agents.

Ultimately, this book has tried to widen the scope for reimagining wetlands as they continue to be lost, by undertaking a multiscale and more-than-human history of these places, showing how they have been shaped with and against other species. The chapters have emphasized some of the diversity of relationships that have cocreated wetlands, which encompass multiple organisms and multiple cultures. People's understandings of wetlands matter. This history shows that shifts in these understandings and relationships are possible. Things could be different.

Interdisciplinary approaches within and across the humanities, social sciences, and sciences are essential in understanding and responding to the continued loss of wetlands and to challenges not only in their management but for those living with and caring for them. Indeed, interdisciplinary approaches are important in reimagining these places in ways that can help create conditions for their flourishing. This book has tried to further strengthen such dialogue. In particular, the more-than-human approach developed here seeks to bring environmental history into closer conversation with interdisciplinary more-than-human and multispecies scholarship in the environmental humanities, as well as to engage with the sciences.

Wetlands have traveled out into the world in multiple ways, connected though human and more-than-human relationships and movement, and through ideas and bureaucracies as well as people and organisms. Whether we like it or not, wetlands come to us. We are all in a relationship with a wetland, most likely with multiple wetlands, and this means that how we understand these places matters. While changes are clearly needed in the way governments and managers approach wetlands, we are all responsible for the future of these places, as we continue to cocreate these more-than-human worlds.

NOTES

FOREWORD

1 Stephen Pyne's *Burning Bush: A Fire History of Australia* was reprinted as part of his related Cycle of Fire subseries.

INTRODUCTION

1 Nancy Langston, *Where Land and Water Meet: A Western Landscape Transformed* (Seattle: University of Washington Press, 2005); Elizabeth Rush, *Rising: Dispatches from the New American Shore* (Minneapolis: Milkweed Editions, 2018); David Biggs, *Quagmire: Nation-Building and Nature in the Mekong Delta* (Seattle: University of Washington Press, 2012); Phillip Garone, *The Fall and Rise of the Wetlands of California's Great Central Valley* (Berkeley: University of California Press, 2011); Robert Wilson, *Seeking Refuge: Birds and Landscapes of the Pacific Flyway* (Seattle: University of Washington Press, 2010); Hugh Prince, *Wetlands of the American Midwest: A Historical Geography of Changing Attitudes* (Chicago: University of Chicago Press, 2008); Ann Vileisis, *Discovering the Unknown Landscape: A History of America's Wetlands* (Washington, DC: Island Press, 1999); David Blackbourn, *The Conquest of Nature: Water, Landscape, and the Making of Modern Germany* (New York: Random House, 2011); Christopher Morris, "Wetland Colonies: Louisiana, Guangzhou, Pondicherry, and Senegal," in *Cultivating the Colonies: Colonial States and Their Environmental Legacies*, ed. Christina Folke Ax, Niels Brimnes, Niklas Thode Jensen, and Karen Oslund (Athens: Ohio University Press, 2011); Sarah R. Hamilton, *Cultivating Nature: The Conservation of a Valencian Working Landscape* (Seattle: University of Washington Press, 2018).
2 See, for example, Ruth Morgan, *Running Out? Water in Western Australia* (Perth: University of Western Australia Press, 2015); P. Davies and S. Lawrence, *Sludge:*

Disaster on Victoria's Goldfields (Melbourne: Black, 2019); Margaret Cook, *A River with a City Problem: A History of Brisbane Floods* (Brisbane: University of Queensland Press, 2019); Deb Anderson, *Endurance: Australian Stories of Drought* (Collingwood, Vic.: CSIRO, 2014); Emily O'Gorman, *Flood Country: An Environmental History of the Murray-Darling Basin* (Collingwood, Vic: CSIRO, 2012).

3 G. Bino, R. T. Kingsford, and K. Brandis, "Australia's Wetlands—Learning from the Past to Manage for the Future," *Pacific Conservation Biology* 22, no. 2 (2016): 117; N. C. Davidson, "How Much Wetland Has the World Lost? Long-term and Recent Trends in Global Wetland Area," *Marine and Freshwater Research* 65 (2014): 940.

4 See, for example, Biggs, *Quagmire*; Garone, *Wetlands*; Wilson, *Seeking Refuge*; Rodney James Giblett, *Postmodern Wetlands: Culture, History, Ecology* (Edinburgh: Edinburgh University Press, 1996); John Charles Ryan and Li Chen, eds., *Australian Wetland Cultures: Swamps and the Environmental Crisis* (London: Lexington Books, 2019).

5 Kevin L. Erwin, "Wetlands and Global Climate Change: The Role of Wetland Restoration in a Changing World," *Wetlands Ecology and Management* 17, no. 1 (2009): 71–84; William J. Mitsch, Blanca Bernal, Amanda M. Nahlik, Ülo Mander, Li Zhang, Christopher J. Anderson, Sven E. Jørgensen, and Hans Brix, "Wetlands, Carbon, and Climate Change," *Landscape Ecology* 28, no. 4 (2013): 583–97; "Briefing Note 10: Wetlands Restoration for Climate Change Resilience," Ramsar Convention on Wetlands of International Importance, https://www.ramsar.org/sites/default/files/documents/library/bn10_restoration_climate_change_e.pdf; "Wetlands and Climate Change," Australian Government Department of Environment and Energy, https://www.environment.gov.au/water/wetlands/publications/wetlands-climate-change.

6 This concept is contested across the sciences, social sciences, and humanities. See, for example, Ruth Morgan, "The Anthropocene as Hydro-Social Cycle: Histories of Water and Technology for the Age of Humans," *Icon* 23 (2017): 36–54; Donna Haraway, "Anthropocene, Capitalocene, Plantationocene, Chthulucene: Making Kin," *Environmental Humanities* 6, no. 1 (2015): 159–65; Jan Zalasiewicz, Colin N. Waters, Mark Williams, and Colin Summerhayes, eds., *The Anthropocene as a Geological Time Unit: A Guide to the Scientific Evidence and Current Debate* (Cambridge: Cambridge University Press, 2019).

7 Bino, Kingsford, and Brandis, "Australia's Wetlands," 117; Daniel Connell, *Water Politics in the Murray-Darling Basin* (Canberra: Federation Press, 2007); O'Gorman, *Flood*. Mounting environmental management challenges in the 1990s underpinned the bureaucratic consolidation of the Murray-Darling Basin as a managerial unit in the Australian federal government. O'Gorman, *Flood*, 220.

8 Bino, Kingsford, and Brandis, "Australia's Wetlands," 117.

9 C. M. Finlayson, Jennifer Ann Davis, Peter A. Gell, R. T. Kingsford, and K. A. Parton, "The Status of Wetlands and the Predicted Effects of Global Climate

Change: The Situation in Australia," *Aquatic Sciences* 75, no. 1 (2013): 73–93; Jamie Pittock and C. Max Finlayson, "Australia's Murray-Darling Basin: Freshwater Ecosystem Conservation Options in an Era of Climate Change," *Marine and Freshwater Research*, 62, no. 3 (2011): 232–43; Jamie Pittock, "The Murray-Darling Basin: Climate Change, Infrastructure, and Water," in *Increasing Resilience to Climate Variability and Change*, ed. Cecilia Tortajada (Singapore: Springer, 2016), 41–59.

10 Bino, Kingsford, and Brandis, "Australia's Wetlands," 117.
11 "Australia's Ramsar Sites," Australian Government Department of Environment and Energy, https://www.environment.gov.au/water/wetlands/australian-wetlands-database/australian-ramsar-wetlands.
12 "Water and the Murray-Darling Basin—a Statistical Profile, 2000–01 to 2005–06," Australian Bureau of Statistics, Australian Government, https://www.abs.gov.au/ausstats/abs@.nsf/mf/4610.0.55.007.
13 Bruce Pascoe, *Dark Emu Black Seeds: Agriculture or Accident?* (Broome, Western Australia: Magabala Books, 2014); Lesley Head, *Second Nature: The History and Implications of Australia as Aboriginal Landscape* (Syracuse: Syracuse University Press, 2000); William B. Gammage, *The Biggest Estate on Earth: How Aborigines Made Australia* (Sydney: Allen and Unwin, 2011).
14 Sue Jackson and Bradley Moggridge, "Indigenous Water Management," *Australasian Journal of Environmental Management* 6, no. 3 (2019): 193–96; Bradley Moggridge, "Where Is the Aboriginal Water Voice through the Current Murray-Darling Crisis?," *Irrigation Australia: The Official Journal of Irrigation Australia* 34 (2018): 34–35; Jessica Weir, *Murray River Country: An Ecological Dialogue with Traditional Owners* (Canberra: Aboriginal Studies Press, 2009).
15 Pascoe, *Dark Emu*.
16 Heather Goodall, "Digging Deeper: Ground Tanks and the Elusive 'Indian Archipelago,'" in *Beyond the Black Stump: Histories of Outback Australia*, ed. Alan Mayne (Adelaide: Wakefield Press, 2008), 129–60; Heather Goodall, "Landscapes of Meaning: The View from Within the Indian Archipelago," *Transforming Cultures eJournal* 3, no. 1 (2008).
17 Richard Kingsford, "Ecological Impacts of Dams, Water Diversions and River Management on Floodplain Wetlands in Australia," *Austral Ecology* 25, no. 2 (2000): 118.
18 O'Gorman, *Flood*.
19 Trevor Langford-Smith and Ian Rutherford, *Water and Land: Two Case Studies in Irrigation* (Canberra: Australian National University Press, 1966), 1.
20 See also Tim Low, *The New Nature* (Sydney: Penguin, 2017).
21 For example, C. J. Lloyd, *Either Drought or Plenty: Water Development and Management in NSW* (Sydney: NSW Department of Water Resources, 1988); Joseph Michael Powell, *Watering the Garden State: Water, Land and Community in Victoria, 1834–1988* (Sydney: Allen and Unwin, 1989).

22 Emily O'Gorman, "Growing Rice on the Murrumbidgee River: Cultures, Politics, and Practices of Food Production and Water Use, 1900 to 2012," *Journal of Australian Studies* 37, no.1 (2013): 96–115.

23 Wilson, *Seeking*.

24 Donna Haraway, "Situated Knowledges: The Science Question in Feminism and the Privilege of Partial Perspective," *Feminist Studies* 14, no. 3 (1988): 575–99. My discussion of more-than-human histories draws on conversations with Andrea Gaynor. See Emily O'Gorman and Andrea Gaynor, "More-Than-Human Histories," *Environmental History* 25, no. 4 (2020), https://doi.org/10.1093/envhis/emaa027.

25 Alfred W. Crosby, *Ecological Imperialism* (Cambridge: Cambridge University Press, 1986); Harriet Ritvo, *The Animal Estate: The English and Other Creatures in the Victorian Age* (Boston: Harvard University Press, 1987); Harriet Ritvo, *Noble Cows and Hybrid Zebras: Essays on Animals and History* (Charlottesville: Virginia University Press, 2010); Susan Nance, ed., *The Historical Animal* (Syracuse, NY: Syracuse University Press, 2015).

26 Hannes Bergthaller, Rob Emmett, Adeline Johns-Putra, Agnes Kneitz, Susanna Lidström, Shane McCorristine, Isabel Pérez Ramos, Dana Phillips, Kate Rigby, and Libby Robin, "Mapping Common Ground: Ecocriticism, Environmental History, and the Environmental Humanities," *Environmental Humanities* 5, no. 1 (2014): 263.

27 Kristin Asdal, "The Problematic Nature of Nature: The Post-constructivist Challenge to Environmental History," *History and Theory* 42, no.4 (2003): 60–74; Mark Carey, "Latin American Environmental History: Current Trends, Interdisciplinary Insights, and Future Directions," *Environmental History* 14, no. 2 (2009): 221–52; Sverker Sörlin and Paul Warde, "The Problem of the Problem of Environmental History: A Re-reading of the Field," *Environmental History* 12, no. 1 (2007): 107–30; Donna Haraway, *When Species Meet* (Minneapolis: University of Minnesota Press, 2008); Bruno Latour, *We Have Never Been Modern* (Cambridge, MA: Harvard University Press, 1993).

28 Paul Sutter, "The World with Us: The State of American Environmental History," *Journal of American History* 100, no. 1 (2013): 94–119.

29 Gregg Mitman, "Living in a Material World," *Journal of American History* 100 (June 2013): 130.

30 See, for example, Etienne Benson, "Animal Writes: Historiography, Disciplinarity, and the Animal Trace," in *Making Animal Meaning*, ed. L. Kalof and G. M. Montgomery (East Lansing: Michigan State University Press, 2011), 3–16; O'Gorman and Gaynor, "More-Than-Human Histories."

31 Thom van Dooren, Eben Kirksey, and Ursula Münster, "Multispecies Studies: Cultivating Arts of Attentiveness," *Environmental Humanities* 8, no. 1 (2016): 1–23.

32 For example, Heather Goodall and Allison Cadzow, *Rivers and Resilience: Aboriginal People on Sydney's Georges River* (Sydney: University of New South Wales Press, 2009).

33 Anna Tsing, "Unruly Edges: Mushrooms as Companion Species for Donna Haraway," *Environmental Humanities* 1, no. 1 (2012): 144.
34 Anna Lowenhaupt Tsing, *The Mushroom at the End of the World: On the Possibility of Life in Capitalist Ruins* (Princeton, NJ: Princeton University Press, 2015), 17–25.
35 Elizabeth A. Grosz, *Volatile Bodies: Toward a Corporeal Feminism* (Bloomington: Indiana University Press, 1994); Astrida Neimanis, *Bodies of Water: Posthuman Feminist Phenomenology* (London: Bloomsbury, 2017).

CHAPTER ONE. WEAVING: POSTCOLONIAL AND MULTISPECIES POLITICS OF PLANTS

1 Richard T. Kingsford and Rachael F. Thomas, "The Macquarie Marshes in Arid Australia and Their Waterbirds: A 50-Year History of Decline," *Environmental Management* 19, no. 6 (1995): 867–78; Richard T. Kingsford and Kristin M. Auld, "Waterbird Breeding and Environmental Flow Management in the Macquarie Marshes, Arid Australia," *River Research and Applications* 21, nos. 2–3 (2005): 187–200.
2 For example, "Pumped: Who's Benefitting from the Billions Spent on the Murray-Darling?," *4 Corners, ABC Television*, posted July 24, 2017, https://www.abc.net.au/4corners/pumped/8727826.
3 Tom Carney, in NSW Department of Environment, Climate Change, and Water, *Aboriginal Cultural Values of the Macquarie Marshes and Gwydir Wetlands: Oral History* (Sydney: Department of Environment, Climate Change and Water, 2010).
4 Kay Masman and Margaret Johnstone, *Reedbed Country: The Story of the Macquarie Marshes* (Tamworth: Macquarie Marshes Catchment Management Committee, 2000), 93–94.
5 Interview with Danielle Carney Flakelar, July 2018; Cameron Muir, *The Broken Promise of Agricultural Progress: An Environmental History* (London: Routledge, 2014), 89–90; Bill Gammage, "Galahs," *Australian Historical Studies* 40, no. 3 (2009): 275–93; Masman and Johnstone, *Reedbed*, 93–94.
6 Timothy J. Ralph, Paul Hesse, and Tsuyoshi Kobayashi, "Wandering Wetlands: Spatial Patterns of Historical Channel and Floodplain Change in the Ramsar-Listed Macquarie Marshes, Australia," *Marine and Freshwater Research* 67, no. 6 (2016): 782–802.
7 See also Northern Basin Aboriginal Nations and Murray-Darling Basin Association with Uncle Neville Merritt, Raymond Thompson, and Danielle Carney Flakelar, "Macquarie Marshes 2015–16, Environmental Watering Event," in Murray-Darling Basin Authority, "Basin Environmental Watering Priorities June 2017: Aboriginal Environmental Outcomes in the Macquarie Marshes," 2017.
8 Interview with Danielle Carney Flakelar, July 2018.

9 Masman and Johnstone, *Reedbed*, 195–227; Muir, *Broken*; O'Gorman, *Flood*, 190–93; Siobhan McHugh, *Cottoning On: Stories of Australian Cotton-Growing* (Sydney: Hale and Iremonger, 1996).
10 Kingsford, "Ecological"; Masman and Johnstone, *Reedbed*, 195–227.
11 Deborah Bird Rose, *Reports from a Wild Country: Ethics for Decolonization* (Sydney: University of New South Wales Press, 2004), 160–71.
12 Interview with Danielle Carney Flakelar, July 2018.
13 Jane Roberts and Frances Marston, *Water Regime for Wetland and Floodplain Plants: A Source Book for the Murray-Darling Basin* (Canberra: National Water Commission, 2011).
14 See also Bawaka Country, including L. Burarrwanga, R. Ganambarr, M. Ganambarr-Stubbs, B. Ganambarr, D. Maymuru, S. Wright, S. Suchet-Pearson, K. Lloyd and J. Sweeney, "Co-becoming Time/s: Time/s-as-telling-as-time/s," in *Methodological Challenges in Nature-Culture and Environmental History Research*, ed. J. Thorpe, S. Rutherford, and L. A. Sandberg (London: Routledge, 2017).
15 Masman and Johnstone, *Reedbed*; Weir, *Murray*; Jessica Weir, D. R. J. Crew, and J. L. Crew, "Wetland Forest Culture: Indigenous Activity for Management Change in the Southern Riverina, New South Wales," *Australasian Journal of Environmental Management* 20, no. 3 (2013): 193–207.
16 On human-plant relationships, see L. Head, J. Atchison, C. Phillips, and K. Buckingham, "Vegetal Politics: Belonging, Practices and Places," *Social and Cultural Geography* 15, no. 8 (2014): 861–70; Sarah Besky and Jonathan Padwe, "Placing Plants in Territory," *Environment and Society* 7, no. 1 (2016): 9–28. My use of the term "postcolonial" refers to the centering of Aboriginal knowledges and a simultaneous decentering of colonial structures and knowledges, while also acknowledging the ongoing legacies of colonialism. The terms "postcolonial" and "decolonizing" are subject to ongoing debate. See, for example, the special issue of *History and Theory* titled "Decolonizing Histories in Theory and Practice" (59, no. 3).
17 Tim Ingold, *The Perception of the Environment: Essays on Livelihood, Dwelling and Skill* (London: Routledge, 2002), 346 and 347.
18 Ingold, *Perception*, 361.
19 See, for example, Lak Lak Burarrwanga, Djawundil Maymuru, Ritjilili Ganambarr, Banbapuy Ganambarr, Sarah Wright, Sandie Suchet-Pearson, and Kate Lloyd, *Weaving Lives Together at Bawaka: North East Arnhem Land* (Callaghan, NSW: University of Newcastle, 2008).
20 Interview with Kate (name changed to protect identity), August 2018. See also Daphne Nash, "'Heritage Knowledge': Indigenous People and Fibre Plants on the NSW South Coast," *Artefact: The Journal of the Archaeological and Anthropological Society of Victoria* 35 (2012): 50–58. For similar revivals, see Ann-Elise Lewallen, *The Fabric of Indigeneity: Ainu Identity, Gender, and Settler Colonialism in Japan* (Albuquerque: University of New Mexico Press, 2016); R. W. Kimmerer, *Braiding Sweetgrass: Indigenous*

Wisdom, Scientific Knowledge and the Teachings of Plants (Minneapolis: Milkweed Editions, 2013); Sarah H. Hill, *Weaving New Worlds: Southeastern Cherokee Women and Their Basketry* (Chapel Hill: University of North Carolina Press, 1997).

21 Nash, "'Heritage'"; Emilie Ens, Fiona Walsh, and Philip Clarke, "Aboriginal People and Australia's Vegetation: Past and Current Interactions," in *Australian Vegetation*, 3rd ed., ed. David A. Keith (Cambridge: Cambridge University Press, 2017), 89–112.

22 Ens, Walsh, and Clarke, "Aboriginal People"; Pascoe, *Dark Emu*; Head, *Second*; Muir, *Broken*; Gammage, *Biggest*; Nash, "'Heritage.'"

23 C. Eriksen and D. L. Hankins, "Colonisation and Fire: Gendered Dimensions of Indigenous Fire Knowledge Retention and Revival," in *The Routledge Handbook of Gender and Development*, ed. A. Coles, L. Gray and J. Momsen (New York: Routledge, 2015), 126.

24 Deborah Bird Rose, "Land Rights and Deep Colonising: The Erasure of Women," *Aboriginal Law Bulletin* 3, no. 85 (1996): 6–13; Diane Bell, *Ngarrindjeri Wurruwarrin: A World That Is, Was, and Will Be* (North Geelong, Vic.: Spinifex Press, 1998); Phyllis Kaberry, *Aboriginal Woman Sacred and Profane* (London: Routledge, 2005).

25 "The Living Murray Program," Murray-Darling Basin Authority, https://www.mdba.gov.au/discover-basin/environment/significant-environmental-sites/icon-sites-along-river-murray; "Information Sheet on Ramsar Wetlands: The Coorong, and Lakes Alexandrina and Albert," Ramsar Convention on Wetlands for International Importance, http://www.environment.gov.au/water/topics/wetlands/database/pubs/25-ris.pdf.

26 "Cyperus gymnocaulos Steud," Global Biodiversity Information Facility, *Atlas of Living Australia*, https://bie.ala.org.au/species/http://id.biodiversity.org.au/node/apni/2915996#.

27 Interview with Meryl and Peter Mansfield-Cameron, March 2017.

28 Interview with Meryl and Peter Mansfield-Cameron, March 2017.

29 Elizabeth Dori Tunstall, "Be Rooted: Learning from Aboriginal Dyeing and Weaving," *The Conversation*, August 11, 2015, https://theconversation.com/be-rooted-learning-from-aboriginal-dyeing-and-weaving-45940.

30 Interview with Danielle Carney Flakelar, July 2018.

31 See also Bawaka Country, Sarah Wright, Sandie Suchet-Pearson, Kate Lloyd, Laklak Burarrwanga, Ritjilili Ganambarr, Merrkiyawuy Ganambarr-Stubbs, Banbapuy Ganambarr, Djawundil Maymuru, and Jill Sweeney, "Co-becoming Bawaka: Towards a Relational Understanding of Place/Space," *Progress in Human Geography* 40, no. 4 (2016): 455–75.

32 Interview with Kate, August 2018.

33 Eriksen and Hankins, "Colonisation"; C. Eriksen, "Gendered Dimensions of Aboriginal Australian and California Indian Fire Knowledge Retention and Revival," *Current Conservation* 7, no.1 (2013): 22–26.

34 Interview with Kate, August 2018.

35 Eriksen and Hankins, "Colonisation."
36 Interview with Danielle Carney Flakelar, July 2018.
37 Interview with Kate, August 2018
38 Interview with Danielle Carney Flakelar, July 2018.
39 Interview with Kate, August 2018.
40 Interview with Danielle Carney Flakelar, July 2018.
41 Interview with Danielle Carney Flakelar, July 2018.
42 Interview with Meryl and Peter Mansfield-Cameron, March 2017.
43 Interview with Danielle Carney Flakelar, July 2018.
44 Pittock, "Murray-Darling."
45 Interview with Kate, August 2018; Finlayson et al., "Status of Wetlands."
46 Billy Griffiths, "Friday Essay: Caring for Country and Telling its Stories," *The Conversation*, May 5, 2017, https://theconversation.com/friday-essay-caring-for-Country-and-telling-its-stories-75844; Ghost Nets Australia, https://www.ghostnets.com.au/.
47 Bell, *Ngarrindjeri*, 75–83; Cameron Muir, "What Do Feather Flowers Tell Us about Our River Systems?," *Everyday Futures*, https://everydayfutures.com.au/project/feather-flowers/; Herbert E. Reid collection, National Museum of Australia, http://collectionsearch.nma.gov.au/.
48 Louise Hamby and Doreen Mellor, "Fibre Tracks," in *Oxford Companion to Aboriginal Art and Culture*, ed. Margo Neale, Sylvia Kleinert, and Robyne Bancroft (Oxford: Oxford University Press, 2000), 371–72; Ellen Trevorrow, Yvonne Koolmatrie, Doreen Kartinyeri, and Diana Wood Conroy, "Binding the Rushes: Survival of Culture," in *Oxford Companion to Aboriginal Art and Culture*, ed. Margo Neale, Sylvia Kleinert, and Robyne Bancroft (Oxford: Oxford University Press, 2000), 371–72.
49 Bell, *Ngarrindjeri*, 75.
50 Muir, "Feather."
51 Hamby and Mellor, "Fibre," 372.
52 Bell, *Ngarrindjeri*, 73, 76, and 78–84.
53 Charles Sturt, *Two Expeditions into the Interior of Southern Australia* (London: Smith, Edler, 1833), 155. See also Bell, *Ngarrindjeri*, 8.
54 Bell, *Ngarrindjeri*, 78–84.
55 "Ellen Trevorrow," Design and Art Australia Online, https://www.daao.org.au/bio/ellen-trevorrow/biography/.
56 Ellen Trevorrow quoted in Diane Bell, ed. *Listen to Ngarrindjeri Women Speaking* (Melbourne: Spinifex Press: 2008), esp. 6 and 7.
57 Interview with Danielle Carney Flakelar, July 2018; Marcia Langton, "Earth, Wind, Fire and Water: The Social and Spiritual Construction of Water in Aboriginal Societies," in *The Social Archaeology of Australian Indigenous Societies*, ed. Bruno David, Bryce Barker, and Ian McNiven (Canberra: Aboriginal Studies Press, 2006), 139–60.
58 Interview with Danielle Carney Flakelar, July 2018.

59 See also Weir, *Murray*.
60 Interview with Danielle Carney Flakelar, July 2018.
61 Bradley J. Moggridge, Lyndal Betteridge, and Ross M. Thompson, "Integrating Aboriginal Cultural Values into Water Planning: A Case Study from New South Wales, Australia," *Australasian Journal of Environmental Management* 26, no. 3 (2019): 273–86; Weir, *Murray*, 87–88, 119–29.
62 Weir, *Murray*, 89; Sue Jackson, Carmel Pollino, Kirsten Maclean, Rosalind Bark, and Bradley Moggridge, "Meeting Indigenous Peoples' Objectives in Environmental Flow Assessments: Case Studies from an Australian Multi-Jurisdictional Water Sharing Initiative," *Journal of Hydrology* 522 (2015): 141–51.
63 Sue Jackson and Deb Nias, "Watering Country: Aboriginal Partnerships with Environmental Water Managers of the Murray-Darling Basin, Australia," *Australasian Journal of Environmental Management* 26, no. 3 (2019): 287–303.
64 Interview with Danielle Carney Flakelar, July 2018. Cultural and environmental flows do not always align with each other. See Jessica McLean, "Still Colonising the Ord River, Northern Australia," *Geographical Journal* 180, no. 3 (2014): 198–210.
65 Weir, *Murray*, 87–89; Sue Jackson and Marcia Langton, "Trends in the Recognition of Indigenous Water Needs in Australian Water Reform: The Limitations of 'Cultural' Entitlements in Achieving Water Equity," *Journal of Water Law* 22, nos. 2–3 (2011): 109–23.
66 Interview with Danielle Carney Flakelar, July 2018. See also Katherine Selena Taylor, Bradley J. Moggridge, and Anne Poelina, "Australian Indigenous Water Policy and the Impacts of the Ever-Changing Political Cycle," *Australasian Journal of Water Resources* 20, no. 2 (2016): 132–47.
67 Moggridge, Betteridge, and Thompson, "Integrating," 5–8.
68 Interview with Danielle Carney Flakelar, July 2018.
69 Interview with Danielle Carney Flakelar, July 2018.
70 Bill Gammage, "The Wiradjuri War, 1838–40," *The Push* 16 (1983): 3–17.
71 Robert Hughes, *The Fatal Shore* (London: Guild, 1986), 276.
72 Masman and Johnstone, *Reedbed*, 71–196.
73 NSW Office of Environment and Heritage, "Macquarie Marshes Adaptive Management Plan," State Government of NSW, June 1, 2010, 64.
74 Masman and Johnstone, *Reedbed*, 25–69; Muir, *Broken*; NSW Office of Environment and Heritage, *Macquarie Marshes*, 67; NSW Department of Environment, Climate Change, and Water, *Aboriginal*, 47–54; Weir, Crew, and Crew, "Wetland," 194; Marcia Langton, "Arts, Wilderness and Terra Nullius," in *Ecopolitics IX: Perspectives on Indigenous People's Management Of Environment Resources*, ed. R. Sultan (Darwin: Northern Land Council, 1995).
75 Interview with Danielle Carney Flakelar, July 2018.
76 Dan Brockington, *Fortress Conservation: The Preservation of the Mkomazi Game Reserve, Tanzania* (Bloomington: Indiana University Press, 2002).

77 Interview with Danielle Carney Flakelar, July 2018.
78 Interview with Danielle Carney Flakelar, July 2018.
79 Interview with Danielle Carney Flakelar, July 2018.

CHAPTER TWO. LEAKING: CONTAINMENT AND RECALCITRANCE OF SWAMPS

Parts of this chapter are adapted from "Swamplands: Human-Animal Relationships in Place," in *Animals Count: How Population Size Matters in Animal-Human Relations*, ed. Nancy Cushing and Jodi Frawley (London: Routledge, 2018), 85–98 (© 2018, Reproduced with permission of the Licensor through PLSclear).

1   For example, Michael Cathcart, *The Water Dreamers: The Remarkable History of Our Dry Continent* (Melbourne: Text Publishing, 2010), 8–38; Grace Karskens, *The Colony: A History of Early Sydney* (Crows Nest, NSW: Allen and Unwin: 2009), 248–51.
2   Mary Douglas, *Purity and Danger: An Analysis of Concepts of Pollution and Taboo* (London: Routledge, [1966] 2003); Grosz, *Volatile*; Thom van Dooren, "Biopatents and the Problem/Promise of Genetic Leaks: Farming Canola in Canada," *Capitalism Nature Socialism* 21, no. 2 (2010): 43–63; Donna Haraway, "The Promises of Monsters: A Regenerative Politics for Inappropriate/d Others," in *Cultural Studies*, ed. Lawrence Grossberg, Cary Nelson, and Paula A Treichler (New York: Routledge, 1992).
3   Grosz, *Volatile*, 194–205.
4   Archaeo, "An Assessment of Cultural Heritage Values Associated with Gowrie Creek Waterways," Toowoomba City Council, 2002, 17–29; Maurice French, *Conflict on the Condamine: Aborigines and the European Invasion* (Toowoomba: Darling Downs Institute Press, 1989), 12–21; Maurice French, *Toowoomba: A Sense of History, 1940-2008* (Toowoomba: University of Southern Queensland, 2009); Paddy Jerome, "*Boobarran Ngummin*: The Bunya Mountains," *Queensland Review* 9, no. 2 (2002): 1–5.
5   Marion Diamond, "The Myth of Patrick Leslie," *Queensland History Journal* 20, no. 11 (2009): 606–17. For example, L. J. Jay, "Pioneer Settlement on the Darling Downs," *Scottish Geographical Magazine* 73, no. 1 (1957): 35–49; Thos. Mathewson, "Personal Reminiscences of '53," *Journal of the Royal Historical Society of Queensland* 1, no. 2 (1916): 96–98.
6   French, *Toowoomba*, 29–39; French, *Conflict*, 39–120.
7   *Queenslander*, 3 July 1869, 3; French, *Toowoomba*, 29–39.
8   French, *Toowoomba*, 34; Peter Cullen, *Toowoomba's Story in Brief* (Toowoomba: Toowoomba Historical Society, 2003), 6.
9   *Toowoomba Chronicle and Darling Downs Gazette*, 21 July 1962.
10 French, *Conflict*, 91–120; R. Kerkhove, "Aboriginal Camps as Urban Foundations? Evidence from Southern Queensland," *Aboriginal History* 42 (2018): 141–72.
11 French, *Toowoomba*, 29–39.

12 Benjamin Glennie Diary, John Oxley Library, OM67-25, SLQ, 23.
13 Nehemiah Bartley, *Opal and Agates; or, Scenes under the Southern Cross and the Magelhans: Being Memories of Fifty Years of Australia and Polynesia* (Brisbane: Gordan and Gotch, 1892), 114.
14 Giblett, *Postmodern*, 1–4; James Beattie, *Empire and Environmental Anxiety: Health, Science, Art and Conservation in South Asia and Australasia, 1800–1920* (London: Springer, 2011), 42–43.
15 Heather Goodall, "Main Streets and Riverbanks: The Politics of Place in an Australian River Town," in *Echoes from the Poisoned Well: Global Memories of Environmental Injustice*, ed. Martin Melosi and Jeffrey Stine (Lanham, MD: Lexington Books, 2006), 73–96.
16 French, *Conflict*, 122–31.
17 French, *Toowoomba*, 2 and 40.
18 *DDGGA*, 26 September 1861, 3; *Courier*, 31 May 1862, 4; Bob Dansie, *A Short History of Gowrie Creek* (Toowoomba: Toowoomba City Council, 1998), 6.
19 Dansie, *Short*, 9.
20 *Courier*, 17 September 1861, 2
21 Christopher Lee, "Spirit of Place: The European Fashioning of Toowoomba," *Queensland Review* 3, no. 1 (1996): 28.
22 Correspondence, briefs, maps, and associated papers relating to the acquisition of swamp reserves at Toowoomba, 1900–1915, Crown Solicitor, 36/3424 (Correspondence), QSA. See also Andrea Gaynor, *Harvest of the Suburbs: An Environmental History of Growing Food in Australian Cities* (Crawley: University of Western Australia Press, 2006), 19.
23 Correspondence, QSA; Warwick Anderson, "Excremental Colonialism: Public Health and the Poetics of Pollution," *Critical Inquiry* 21, no. 3 (1995): 644; Beattie, *Empire*, 42–43; Stephen Halliday, "Death and Miasma in Victorian London: An Obstinate Belief," *BMJ: British Medical Journal* 323, no. 7327 (2001): 1469–71.
24 *DDGGA*, 1 July 1865, 3.
25 Halliday, "Death"; Beattie, *Empire*, 42–43; Melanie A. Kiechle, *Smell Detectives: An Olfactory History of Nineteenth-Century Urban America* (Seattle: University of Washington Press, 2017), 5–6.
26 Anderson, "Excremental," 644; Beattie, *Empire*, 42–43; Halliday, "Death."
27 Correspondence, QSA.
28 W. K. Hancock, *Discovering Monaro* (Cambridge: Cambridge University Press, 1972), 3.
29 Tom Griffiths, *Forests of Ash: An Environmental History* (Cambridge: Cambridge University Press, 2001), 32, 40–42. See also Tracey Banivanua Mar, "Carving Wilderness: Queensland's National Parks and the Unsettling of Emptied Lands, 1890–1910," in *Making Settler Colonial Space: Perspectives on Race, Place and Identity*, ed. Banivanua Mar and Penny Edmonds (Basingstoke: Palgrave, 2010), 73–94.
30 Griffiths, *Forests of Ash*, 32–35.

31 Correspondence, QSA.
32 Correspondence, QSA.
33 Correspondence, QSA.
34 Correspondence, QSA. For example, Vera S. Candiani, *Dreaming of Dry Land: Environmental Transformation in Colonial Mexico City* (Stanford, CA: Stanford University Press, 2014).
35 Correspondence, QSA.
36 Correspondence, QSA.
37 *Toowoomba Chronicle and Queensland Advertiser*, 4 April 1874, 2.
38 *Toowoomba Chronicle*, 26 February 1876, 5; Correspondence, QSA.
39 *Toowoomba Chronicle*, 18 July 1876, 3; *DDGGA*, 22 July 1876, 2.
40 Halliday, "Death," 1470–71; Stephen Halliday *The Great Stink of London: Sir Joseph Bazalgete and the Cleansing of the Victorian Capital* (Stroud: Sutton, 1999); A. S. Wohl, *Endangered Lives: Public Health in Victorian Britain* (London: Methuen, 1984).
41 Margaret Pelling, "Contagion/Germ Theory/Specificity," in *Companion Encyclopedia of the History of Medicine*, vol. 1, ed. W. F. Bynam and R. Porter (London: Routledge, 1993), 309–10. Germ theory began to gain prominence in Europe in the mid to late nineteenth century but has a longer history.
42 *TCDDGA*, 11 May 1878, 3.
43 *TCDDGA*, 11 May 1878, 3; Donna Haraway, *Modest_Witness@Second_Millennium. FemaleMan_Meets_OncoMouse: Feminism and Technoscience* (Brighton, UK: Psychology Press, 1997), 28; Steven Shapin and Simon Schaffer, *Leviathan and the Air-Pump: Hobbes, Boyle, and the Experimental Life* (Princeton, NJ: Princeton University Press, 2011).
44 *TCDDGA*, 11 May 1878, 3.
45 James Alfred Yelling, *Slums and Slum Clearance in Victorian London* (London: Routledge, 2012).
46 *TCDDGA*, 11 May 1878, 3.
47 Joanna Boileau, *Chinese Market Gardening in Australia and New Zealand: Gardens of Prosperity* (London: Palgrave Macmillan, 2017), 192–93; *DDG*, 4 April 1894, 3; *DDG*, 29 September 1894, 5; *DDG*, 5 September 1894, 4.
48 Catherine McNeur, "The 'Swinish Multitude': Controversies over Hogs in Antebellum New York City," *Journal of Urban History* 37, no. 5 (2011): 639–60.
49 Anne Fisher, "The Forgotten Pioneers: The Chinese in Southern Inland Queensland (1848 to circa 1914)" (MA diss., University of Southern Queensland, 1995), 38–52; Boileau, *Chinese*, 11; James Beattie, "Dragons Abroad: Chinese Migration and Environmental Change in Australasia," in *Visions of Australia: Environments in History*, ed. Christof Mauch, Ruth Morgan and Emily O'Gorman (Munich: RCC Perspectives, 2017), 59–70; James Beattie, "Chinese Resource Frontiers, Environmental Change, and Entrepreneurship in the Pacific, 1790s–1920s," in *Migrant Ecologies in the Pacific*, ed. Edward Melillo and Ryan Jones (Honolulu: Hawai'i University Press, in press).

50 *DDGGA*, 15 January 1870, 3; Fisher, "Forgotten," 38-52; Boileau, *Chinese*, 5 and 60; Beattie, "Dragons."
51 Boileau, *Chinese*, 11. See also, for example, *DDGGA*, 15 January 1870, 3.
52 For example, *Queensland Times*, 16 February 1893, 6; *Queensland Times*, 25 February 1893, 2.
53 Boileau, *Chinese*, 192-95; Xiaotang Ju, Fusuo Zhang, Xuemei Bao, V. Römheld, and M. Roelcke, "Utilization and Management of Organic Wastes in Chinese Agriculture: Past, Present and Perspectives," *Science in China Series C: Life Sciences* 48, no. 2 (2005): 965-79; Fei Sheng, "Environmental Experiences of Chinese People in the Mid-Nineteenth Century Australian Gold Rushes," *Global Environment* 4, nos. 7-8 (2011): 98-117, 121.
54 *TCDDGA*, 11 May 1878, 3.
55 Raymond Evans, Kay Saunders, and Kathryn Cronin, *Race Relations in Colonial Queensland: A History of Exclusion, Exploitation and Extermination*, 3rd ed. (St. Lucia: University of Queensland Press, 1993), 235-318.
56 Anderson, "Excremental," 643.
57 *TCDDGA*, 11 May 1878, 4.
58 Lee, *Spirit*.
59 *TCDDGA*, 11 May 1878, 3.
60 *DDGGA*, 21 July 1880, 3; Dansie, *Short*, 19.]
61 *DDG*, 27 November 1895, 5.
62 *DDG*, 4 April 1894, 3; *DDG*, 29 September 1894, 5; *DDG*, 5 September 1894, 4.
63 *Warwick Examiner and Times*, 22 February 1893, 3; *The Queenslander*, 4 March 1893, 422; *SMH*, 2 April 1906, 8; *Morning Post*, 3 April 1906, 3.
64 Rosalind Kidd, *The Way We Civilise: Aboriginal Affairs, the Untold Story* (St. Lucia: University of Queensland Press, 1997); Shirleene Robinson and Jessica Paten, "The Question of Genocide and Indigenous Child Removal: The Colonial Australian Context," *Journal of Genocide Research* 10, no. 4 (2008): 501-18; M. Copland, J. Richards, and A. Walker, *One Hour More Daylight: A Historical Overview of Aboriginal Dispossession in Southern and South West Queensland* (Toowoomba: Social Justice Commission, 2010).
65 Jerome, *"Boobarran,"* 3.
66 *DDG*, 3 April 1895, 5; *The Telegraph*, 14 September 1887, 3; Fisher, "Forgotten," 44-47; Evans, Saunders, and Cronin, *Race*, 235-318.
67 For example, *DDG*, 4 April 1894, 3; *DDG*, 29 September 1894, 5; *DDG*, 5 September 1894, 4; *DDG*, 3 April 1895, 5; *DDG*, 18 November 1896, 5.
68 *DDG*, 19 October 1904, 2.
69 *Warwick Daily News*, 1 December 1922, 5; Bob Dansie, "Dr Thomas Price—the Mosquito King," unpublished, Toowoomba Historical Society Collection.
70 Peter H. Curson and Kevin McCracken, *Plague in Sydney: The Anatomy of an Epidemic* (Sydney: University of New South Wales Press, 1989); J. Burton Cleland, Burton Bradley, and W. McDonald, "Dengue Fever in Australia," *Epidemiology and Infection* 16, no. 4 (1918): 317-418.

71 Jeffrey R. Powell, Andrea Gloria-Soria, and Panayiota Kotsakiozi, "Recent History of Aedes Aegypti: Vector Genomics and Epidemiology Records," *Bioscience* 68, no. 11 (2018): 854–60.
72 Rat and Mosquito Destruction, Toowoomba City Council, 1918–1949, Box 12, Books 3 and 4, 18980 (Rat), QSA.
73 Rat, QSA.
74 Anna Tsing, "Others without History: Organisms as Agility-Shifting Actors in the Trajectory of Capital," Eric Wolf lecture, Austrian Academy of Sciences, October 2016.
75 "Men's Toilet, Russell Street, Toowoomba (entry 601381)," Queensland Heritage Register, Queensland Heritage Council.
76 Rat, QSA.
77 *Warwick Daily News*, 1 December 1922, 5; Bob Dansie, "Dr."
78 Rat, QSA; *DDG*, 26 August 1920, 3; Dansie, "Dr."
79 Rat, QSA.
80 *Queensland Government Gazette*, 30 April 1932, 1643; Rat, QSA.
81 Rat, QSA.
82 *DDG*, 26 August 1920, 3.
83 Rat, QSA.
84 Rat, QSA.
85 Rat, QSA.
86 Interview with field naturalist, September 2017. See also Archaeo, "Assessment."
87 Interview with field naturalist, September 2017.

CHAPTER THREE. INFECTING: IRRIGATION, MOSQUITOES, AND MALARIA IN WARTIME

This chapter previously appeared as "Imagined Ecologies: A More-Than-Human History of Malaria in the Murrumbidgee Irrigation Area, New South Wales, Australia, 1919–45," *Environmental History* 22, no. 3 (2017): 486–514.

1 Leo Barney Slater, *War and Disease: Biomedical Research on Malaria in the Twentieth Century* (New Brunswick, NJ: Rutgers University Press, 2009); Beattie, *Empire*, 42–43.
2 *Narrandera Argus and Riverina Herald*, 11 January 1929, 1; *Maitland Mercury*, 17 January 1929, 2; *Daily Advertiser*, 5 January 1946, 6.
3 "Malaria Mosquitoes—Murrumbidgee Irrigation Area," 1917–46, Item 45/7990, Series 14511, Murrumbidgee Irrigation Area [MIA] files [Water Resources Commission] 1911–86 (henceforth "Malaria"), SRNSW. These investigations did not differentiate between different malaria *Plasmodium*.
4 Tess Lea, "Mosquito Architecture," unpublished paper; J. R. McNeill, *Mosquito Empires: Ecology and War in the Greater Caribbean, 1620-1914* (Cambridge: Cambridge University Press, 2010); Timothy Mitchell, *Rule of*

Experts: Egypt, Techno-Politics, Modernity (Berkeley: University of California Press, 2002).
5 Haraway, "Situated."
6 Tsing, "Unruly," 144.
7 For example, Warwick Anderson, "Postcolonial Ecologies of Parasite and Host: Making Parasitism Cosmopolitan," *Journal of the History of Biology* 49, no. 2 (2016): 241–59; J. Andrew Mendelsohn, "From Eradication to Equilibrium: How Epidemics Became Complex after World War I," in *Greater Than the Parts: Holism in Biomedicine, 1920–1950*, ed. Christopher Lawrence and George Weisz (Oxford: Oxford University Press, 1998), 303–31.
8 Warwick Anderson, "Natural Histories of Infectious Disease: Ecological Vision in Twentieth-Century Biomedical Science," *Osiris* 19 (2004): 42. See also Warwick Anderson, "Nowhere to Run, Rabbit: The Cold-War Calculus of Disease Ecology," *History and Philosophy of the Life Sciences* 39, no. 13 (2017): https://doi.org/10.1007/s40656-017-0140-7; Warwick Anderson, *Colonial Pathologies: American Tropical Medicine, Race, and Hygiene in the Philippines* (Durham, NC: Duke University Press, 2006), 213–23.
9 Linda Nash, *Inescapable Ecologies: A History of Environment, Disease, and Knowledge* (Berkeley: University of California Press, 2006).
10 Warwick Anderson, *The Cultivation of Whiteness: Science, Health, and Racial Destiny in Australia* (Durham, NC: Duke University Press, 2006), 2.
11 Beattie, *Empire*, 51–54.
12 Lea, "Mosquito."
13 Tsing, "Unruly," 144, original emphasis.
14 Mary Louise Pratt, "Arts of the Contact Zone," *Profession* (1991), 33–40; Haraway, *Species*.
15 Rod Nan Tie, "Malaria in Australia: Current Trends and Future Risks," *Annals of the ACTM: An International Journal of Tropical and Travel Medicine* 10, no. 2 (2009): 34–37; John Walker, "Malaria in a Changing World: An Australian Perspective," *International Journal for Parasitology* 28, no. 6 (1998): 947–53; Lea, "Mosquito."
16 Robert H. Black, *Malaria in Australia* (Canberra: Australian Government Publishing Service, 1972).
17 Francis E. G. Cox, "History of the Discovery of the Malaria Parasites and Their Vectors," *Parasites and Vectors* 3, no. 5 (2010), DOI: 10.1186/1756-3305-3-5; Gordon Harrison, *Mosquitoes, Malaria and Man: A History of the Hostilities since 1880* (London: John Murray, 1978); Slater, *War*; Randall M. Packard, *The Making of a Tropical Disease: A Short History of Malaria* (Baltimore: John Hopkins University Press, 2007).
18 F. H. Taylor, *Malaria Mosquito Survey* (Melbourne: Commonwealth Government of Australia, 1917), in "Malaria," SRNSW; Robert Fowler, "The Risk of Malaria in Australia," *Medical Journal of Australia*, August 2, 1919, 83–86; Walter Summons, "Incidence of Malaria amongst Troops on a Transport to

Australia from Egypt and Palestine," *Medical Journal of Australia*, 2 August 1919, 86–88.
19 Bernard J. Brabin, "Malaria's Contribution to World War One—the Unexpected Adversary," *Malaria Journal* 13, no. 1 (2014): 1–22.
20 Ian Howie-Willis, "Malariology in Australia between the First and Second World Wars" (part 2 of "Pioneers of Australian Military Malariology"), *Journal of Military and Veterans' Health* 24, no. 2 (1 April 2016), http://jmvh.org/article/malariology-in-australia-between-the-first-and-second-world-wars-part-2-of-pioneers-of-australian-military-malariology/.
21 Taylor, *Malaria*; *Jerilderie Herald and Urana Advertiser*, 20 July 1917, 1.
22 Langford-Smith and Rutherford, *Water*, 1.
23 See O'Gorman, *Flood*, esp. 119–34.
24 J. H. L. Cumpston, "Introduction," in Taylor, *Malaria*; Michael Roe, "Cumpston, John Howard Lidgett (1880–1954)," in *Australian Dictionary of Biography*, National Centre of Biography, Australian National University, http://adb.anu.edu.au/biography/cumpston-john-howard-lidgett-5846/text9935 (published first in hardcopy 1981).
25 Brabin, "Malaria's," 1–22; Anderson, *Colonial*, 209–19.
26 *Argus*, 27 July 1918, 16.
27 R. W. Cilento, *Malaria, with Especial Reference to Australia and Its Dependencies* (Melbourne: Australia Department of Health, 1924), 27–28.
28 Howie-Willis, "Malariology"; Cilento, *Malaria*.
29 Anderson, *Cultivation*; Russell McGregor, *Environment, Race and Nationhood in Australia: Revisiting the Empty North* (New York: Palgrave Macmillan, 2016), 10–11; Warwick Frost, "Australia Unlimited? Environmental Debate in the Age of Catastrophe, 1910–1939," *Environment and History* (2004): 285–303.
30 "The Key Climate Groups," Australian Bureau of Meteorology, http://www.bom.gov.au/iwk/climate_zones/map_1.shtml. See also Margaret Humphreys, *Malaria: Poverty, Race, and Public Health in the United States* (Baltimore: Johns Hopkins University Press, 2001).
31 "Malaria," SRNSW.
32 "Malaria," SRNSW.
33 *Riverine Grazier*, 7 November 1916, 1
34 *Riverine Grazier*, 7 November 1916, 1.
35 Lloyd, *Either*, 199–212; Langford-Smith and Rutherford, *Water*, 25–29 and 33–43; Richard Waterhouse, *The Vision Splendid: A Social and Cultural History of Rural Australia* (Fremantle, WA: Curtin University Press), 60–66, 200–201.
36 J. L. Trefle Minister for Agriculture quoted in Lloyd, *Either*, 205. See also Langford-Smith and Rutherford, *Water*, 46–49.
37 Langford-Smith and Rutherford, *Water*, 55.
38 "Malaria," SRNSW.
39 R. V. Southcott, "Cleland, Sir John Burton (1878–1971)," in *Australian Dictionary of Biography*, National Centre of Biography, Australian National University,

http://adb.anu.edu.au/biography/cleland-sir-john-burton-5679/text9595 (published first in hardcopy 1981); Bryan Egan, "Morgan, Frederick Grantley (1891–1969)," in *Australian Dictionary of Biography*, http://adb.anu.edu.au /biography/morgan-frederick-grantley-11164/text19889 (published first in hardcopy 2000).

40 "Malaria," SRNSW.
41 "Malaria," SRNSW.
42 Katrina Saltzman, Lesley Head, and Marie Stenseke, "Do Cows Belong in Nature? The Cultural Basis of Agriculture in Sweden and Australia," *Journal of Rural Studies* 27, no. 1 (2011): 54–62.
43 "Malaria," SRNSW.
44 "Malaria," SRNSW.
45 Peter Vallee, "Ferguson, Eustace William (1884–1927)," in *Australian Dictionary of Biography*, National Centre of Biography, Australian National University, http://adb.anu.edu.au/biography/ferguson-eustace-william-6157/text10575 (published first in hardcopy 1981); Howie-Willis. "Malariology."
46 "Malaria," SRNSW.
47 "Malaria," SRNSW.
48 Lloyd, *Either*, 199–212; Langford-Smith and Rutherford, *Water*, 25–29 and 33–43.
49 W. R. Watkins, "Rice-Growing: Its Possibilities on the Murrumbidgee Irrigation Areas," *Agricultural Gazette of New South Wales*, October 1926, 748; *Growing Rice in New South Wales* (Sydney: NSW Government Printer, c. 1948), 2; "Agriculture—Division of Marketing and Economics, Correspondence, 1923–73," 200, 10/25451, 384A (henceforth "Agriculture"), SRNSW.
50 Lloyd, *Either*, 217.
51 "Agriculture," SRNSW.
52 Emily O'Gorman, "Growing Rice on the Murrumbidgee River: Cultures, Politics, and Practices of Food Production and Water Use, 1900 to 2012," *Journal of Australian Studies* 37, no. 1 (2013): 96–115.
53 Watkins, "Rice-Growing," 748; "Integration of Animal and Crop Production," 200, 10/35499, 68/1132, SRNSW.
54 *Rice: The World's Main Food Crop*, Royal Easter show pamphlet (Sydney: NSW Department of Agriculture, 1965), 7; Watkins, "Rice-Growing," 747.
55 *SMH*, 13 July 1932, 9.
56 Anderson, *Cultivation*, esp. 97–122. See also Watkins, "Rice-Growing," 741.
57 This may have been the mosquito-borne Ross River virus. See Susan Jacups, Peter I. Whelan, and Bart J. Currie, "Ross River Virus and Barmah Forest Virus Infections: A Review of History, Ecology, and Predictive Models, with Implications for Tropical Northern Australia," *Vector-Borne and Zoonotic Diseases* 8, no. 2 (2008): 283–98.
58 "Malaria," SRNSW.
59 E. N. Marks, "Hamlyn-Harris, Ronald (1874–1953)," in *Australian Dictionary of Biography*, National Centre of Biography, Australian National University,

http://adb.anu.edu.au/biography/hamlyn-harris-ronald-6541/text11239 (published first in hardcopy 1983).

60 Anderson, *Cultivation*, 138 and 209.
61 "Malaria," SRNSW.
62 Paul S. Sutter, "Nature's Agents or Agents of Empire? Entomological Workers and Environmental Change during the Construction of the Panama Canal," *Isis* 98, no. 4 (2007): 725.
63 Anderson, "Excremental," 640–69; Jamie Lorimer, "Gut Buddies: Multispecies Studies and the Microbiome," *Environmental Humanities* 8, no. 1 (2016): 57–76.
64 R. Pesman and C. Kevin, *A History of Italian Settlement in New South Wales* (Sydney: NSW Heritage Office, 2001); Edmund Russell, *War and Nature: Fighting Humans and Insects with Chemicals from World War I to Silent Spring* (Cambridge: Cambridge University Press, 2001).
65 *Daily Telegraph*, 16 January 1929, in "Malaria," SRNSW.
66 "Malaria," SRNSW.
67 "Malaria," SRNSW.
68 *Albury Banner and Wodonga Express*, 8 February 1929, 11; "Malaria," SRNSW.
69 "Malaria," SRNSW.
70 Ronald Hamlyn-Harris, "The Consideration of Certain Factors as Potentialities in Mosquito Control in Australia," *Proceedings of the Royal Society of Queensland* 42, no. 10 (1930): 86–105.
71 "Malaria," SRNSW. See also A. White and G. Pyke, "World War II and the Rise of the Plague Minnow Gambusia Holbrooki (Girard, 1859) in Australia," *Australian Zoologist* 35, no. 4 (2011): 1024–32.
72 "Agriculture," SRNSW; James C. Scott, "An Approach to the Problems of Food Supply in Southeast Asia during World War Two," in *Agriculture and Food Supply in the Second World War*, ed. Bernd Martin and Alan S. Milward (St. Katharinen, Germany: Scripta Mercaturae Verlag, 1985), 275–81; O'Gorman. "Growing."
73 *Rice: The World's*, 8; Marnie Haig-Muir, "The Wakool Wartime Rice-Growing Project and Its Impact on Regional Development," *Australian Economic History Review* 36, no. 2 (1996): 66–69.
74 "Malaria," SRNSW.
75 "Malaria," SRNSW.
76 "Malaria," SRNSW.
77 Slater, *War*, 32.
78 Original emphasis. "Malaria," SRNSW. See J. R. McNeill, *Something New under the Sun: An Environmental History of the Twentieth-Century World* (New York: W. W. Norton, 2001); Rachel Carson, *Silent Spring* (1962; repr. Boston: Houghton Mifflin Harcourt, 2002).
79 Dawn Day Biehler and Gregory L. Simon, "The Great Indoors: Research Frontiers on Indoor Environments as Active Political-Ecological Spaces," *Progress in Human Geography* 35, no. 2 (2011): 172–92.
80 "Malaria," SRNSW.

81 "Malaria," SRNSW.
82 F. H. S. Roberts, "Observations on Anopheles Annulipes Walk, as a Possible Vector of Malaria," *Australian Journal of Experimental Biology and Medical Science* 21, no. 4 (1943): 259 and 261. See also Beverley M. Angus, "Roberts, Frederick Hugh Sherston (1901–1972)," in *Australian Dictionary of Biography*, National Centre of Biography, Australian National University, http://adb.anu.edu.au/biography/roberts-frederick-hugh-sherston-11536/text20581 (published first in hardcopy 2002).
83 "Malaria," SRNSW.
84 *Daily Advertiser*, 5 January 1946, 6; "Malaria," SRNSW.
85 Anderson, "Nowhere."
86 Ben D. Ewald, Cameron E. Webb, David N. Durrheim, and Richard C. Russell, "Is There a Risk of Malaria Transmission in NSW?," *New South Wales Public Health Bulletin* 19, no. 8 (2008): 127–31.
87 Cameron Webb, personal correspondence, 9 November 2016; Ewald et al., "Risk"; Richard C. Russell, "Malaria," Department of Medical Entomology, University of Sydney http://medent.usyd.edu.au/fact/malaria.htm; Cameron Webb, Richard Russell, and Stephen Doggett, *A Guide to Mosquitoes of Australia* (Collingwood, Vic.: CSIRO, 2016), 140; D. H. Foley and J. H. Bryan, "Anopheles annulipes Walker (Diptera: Culicidae) at Griffith, New South Wales: 1. Two Sibling Species in Sympatry," *Australian Journal of Entomology* 30, no. 2 (1991): 109–12; John Walker, "Malaria in a Changing World: An Australian Perspective," *International Journal for Parasitology* 28, no. 6 (1998): 947–53; Roberts, "Observations," 259.
88 Ewald et al., "Risk"; Cameron Webb, personal correspondence, 9 November 2016.
89 Haraway, "Situated."

CHAPTER FOUR. CROSSING: WILDLIFE IN AGRICULTURE

Parts of this chapter are adapted from "Remaking Wetlands: Rice Fields and Ducks in the Murrumbidgee River Region, NSW," in *Rethinking Invasion Ecologies from the Environmental Humanities*, ed. Jodi Frawley and Iain McCalman (London: Routledge, 2014), 215–38 (© 2014, Reproduced with permission of The Licensor through PLSclear).

1 Interview with anonymous farmer 1, May 2012.
2 Weir, *Murray*, 26–46; Paul Sinclair, *The Murray: A River and Its People* (Melbourne: Melbourne University Press, 2001), 3–25.
3 For example, Charis Thompson, "When Elephants Stand for Competing Philosophies of Nature: Amboseli National Park, Kenya," in *Complexities: Social Studies of Knowledge Practices*, ed. J. Law and A. Mol (Durham: Duke University Press, 2002): 166–90; Mark Cioc, *The Game of Conservation: International Treaties to Protect the World's Migratory Animals* (Athens: Ohio University Press, 2009); Etienne Benson, *Wired Wilderness: Technologies of*

Tracking and the Making of Modern Wildlife (Baltimore: John Hopkins University Press, 2010).
4 David Roshier, "Grey Teal: Survivors in a Changing World," in *Boom and Bust: Bird Stories for a Dry Country*, ed. Libby Robin, Robert Heinsohn, and Leo Joseph (Collingwood, Vic.: CSIRO, 2009), 76–78; Wilson, *Seeking*.
5 Thom van Dooren and Deborah Bird Rose, "Storied-Places in a Multispecies City," *Humanimalia* 3, no. 2 (2012): 1–27.
6 "Fivebough and Tuckerbil Swamps," Australian Wetlands Database, Ramsar wetlands, Department of Environment and Energy, Australian Government, http://www.environment.gov.au/cgi-bin/wetlands/ramsardetails.pl?refcode=62.
7 O'Gorman, "Growing."
8 Lloyd, *Either*, 199–212; Langford-Smith and Rutherford, *Water*, 25–29 and 33–43.
9 Fivebough and Tuckerbil Management Trust, "Management Plan for Fivebough and Tuckerbil Swamps," September 2002, 15.
10 Lloyd, *Either*, 199 and 209–10; Langford-Smith and Rutherford, *Water*, 25–29.
11 Gammage, "Wiradjuri," 15; Peter Rimas Kabaila, *Wiradjuri Places: The Murrumbidgee River Basin with a Section on Ngunawal Country* (Jamison Centre: Black Mountain Projects, 1995), 117–41.
12 *SMH*, 12 August 1939, 12; Fivebough and Tuckerbil, "Management Plan," 11; Mike Schultz, "Fivebough and Tuckerbil Swamps," *Wetlands Australia* 12 (2004): 10.
13 *SMH*, 12 August 1939, 12.
14 Fivebough and Tuckerbil, "Management Plan," 11.
15 H. J. Frith, "Wild Ducks and the Rice Industry in New South Wales," *Wildlife Research* 2, no. 1 (1957): 41.
16 Thomas R. Dunlap, *Nature and the English Diaspora: Environment and History in the United States, Canada, Australia, and New Zealand* (Cambridge: Cambridge University Press, 1999), 247 and 254; Gary Lewis, *An Illustrated History of the Riverina Rice Industry* (Leeton, NSW: Ricegrowers' Co-operative, 1994), 174–88.
17 I. R. Taylor and A. Richardson, *The Ecology and Management of Waterbirds on Fivebough Swamp*, Johnstone Centre Report No. 141 (Albury, NSW: Charles Sturt University, 2002).
18 Fivebough and Tuckerbil, "Management Plan," 9 and 26–27; H. Glazebrook and I. R. Taylor, *Fivebough and Tuckerbil Swamps: A Review of Their History, Conservation Values and Future Management Options*, Johnstone Centre Report No. 105 (Albury, NSW: Charles Sturt University, 1998).
19 Mike Shultz and Bill Phillips, "Fivebough and Tuckerbil Swamps," Ramsar information sheet, October 2002.
20 Interview with Jamie Pittock and Cath Webb, November 2018.
21 Interview with Jamie Pittock and Cath Webb, November 2018.
22 Interview with Jamie Pittock and Cath Webb, November 2018; Fivebough and Tuckerbil, "Management Plan."

23 "Environment Committee" and "Press Cuttings 1," Ricegrowers' Association of Australia Archives. These archives document critiques and responses to them by the Ricegrowers' Association of Australia. See also Adam Barclay, "A Sunburned Grain," *Rice Today*, April–June 2010, 9, 12–17.
24 R. Kingsford and R. Thomas, "Destruction of Wetlands and Waterbird Populations by Dams and Irrigation on the Murrumbidgee River in Arid Australia," *Environmental Management* 34, no. 3 (2004): 386–94.
25 Richard Kingsford, Grahame Webb, and Peter J. Fullagar, "Scientific Panel Review of Open Seasons for Waterfowl in New South Wales," NSW National Parks and Wildlife Service, November 2000, 4; A. L. Curtin and R. T. Kingsford, "An Analysis of the Problem of Ducks on Rice in New South Wales," National Parks and Wildlife Service, Hurstville, November 1997, 32.
26 Roshier, "Grey Teal," 76–78.
27 Curtin and Kingsford, "Ducks," 3; J. R. Kinghorn, "Wild Ducks Are Not a Serious Pest of Rice Crops," *Agricultural Gazette of New South Wales*, 1 August 1932, 603; Frith, "Wild," 33.
28 Curtin and Kingsford, "Ducks," 9; R. Kingsford and I. Norman, "Australian Waterbirds: Products of the Continent's Ecology," *Emu* 102 (2002): 1–23.
29 Paul Humphries, "Historical Indigenous use of Aquatic Resources in Australia's Murray-Darling Basin, and Its Implications for River Management," *Ecological Management and Restoration*, 8, no. 2 (2007): 107.
30 Curtin and Kingsford, "Ducks," 31.
31 See George Main, *Heartland: The Regeneration of Rural Place* (Sydney: University of New South Wales Press, 2005), 16–56.
32 Curtin and Kingsford, "Ducks," 9–10.
33 Wilson, *Seeking*, 3–15.
34 Emily O'Gorman and Thom van Dooren, "The Promises of Pests: Wildlife in Agricultural Systems," *Australian Zoologist* 39, no. 1 (2017): 81–84.
35 O'Gorman and van Dooren, "Promises."
36 Kinghorn, "Wild," 603.
37 "Puddle" means to muddy and otherwise disturb flooded bays while paddling.
38 Kinghorn, "Wild," 603. See also N. S. Ellis, "Ducks and the Rice Industry," *Emu* 39, no. 3 (1940): 201–2; Correspondence files, C.80/26, 1926, Australian Museum Archives.
39 Frith, "Wild," 33; Correspondence files, C.80/26, 1926, Australian Museum Archives; K. McKeown, "List of Birds of the Murrumbidgee Irrigation Areas," *Emu* 23, no. 1 (1923): 42–48, 43.
40 "Bibliography—Reports on Fauna to Government Departments by J. R. Kinghorn," AMS402, Papers of J. K. (Roy) Kinghorn, 1920–64, Australian Museum Archives.
41 Ellis, "Ducks," 201.
42 Kinghorn, "Wild," 603.
43 Kinghorn, "Wild," 604–6.
44 Kinghorn, "Wild," 604, 606, and 607.

45 Kinghorn, "Wild," 607–8.
46 Kinghorn, "Wild," 607–8.
47 Libby Robin, "Ecology: A Science of Empire?," in *Ecology and Empire: Environmental History of Settler Societies*, ed. Tom Griffiths and Libby Robin (Melbourne: Melbourne University Press, 1997), 68–70; Thomas R. Dunlap, "Ecology and Environmentalism in the Anglo Settler Colonies," in *Ecology and Empire*, ed. Griffiths and Robin, 76–78.
48 Dunlap, "Ecology," 77–78.
49 Kinghorn, "Wild," 606.
50 Kinghorn, unpublished report, quoted in Ellis, "Ducks," 203.
51 Frith, "Wild," 32. The area under rice cultivation in 1931–32 and 1932–33 was approximately twenty thousand acres. "Agriculture," SRNSW.
52 For example, *Sidney Morning Herald*, 9 December 1938, 3. See Deborah Bird Rose, *Reports from a Wild Country: Ethics for Decolonisation* (Sydney: University of New South Wales Press, 2004), 53–72; Deborah Rose, "What If the Angel of History Were a Dog?," *Cultural Studies Review* 12, no. 1 (2006): 67–78; Thom van Dooren, "Invasive Species in Penguin Worlds: An Ethical Taxonomy of Killing for Conservation," *Conservation and Society* 9, no. 4 (2011): 286–98, esp. 290.
53 Introduced animals were those considered to have arrived in Australia after British colonial settlement in 1788. Peter Jarman and Margaret Brock, "The Evolving Intent and Coverage of Legislation to Protect Biodiversity in New South Wales," in *Threatened Species Legislation: Is It Just an Act?*, ed. Pat Hutchings, Daniel Lunney, and Chris Dickman (Mosman: Royal Zoological Society of NSW, 2004), 4; B. J. Stubbs, "From 'Useless Brutes' to National Treasures: A Century of Evolving Attitudes towards Native Fauna in New South Wales, 1860s to 1960s," *Environment and History* 7 (2001): 29–30.
54 R. B. Walker, "Fauna and Flora Protection in New South Wales, 1866–1948," *Journal of Australian Studies* 15 (1991): 18; Jarman and Brock, "Evolving," 2; Stubbs, "'Useless Brutes,'" 26–28; Coral Dow, "'Sportsman's Paradise': The Effects of Hunting on the Avifauna of the Gippsland Lakes," *Environment and History* 14 (2008): 148; Tim Bonyhady, *The Colonial Earth* (Melbourne: University of Melbourne Press, 2000), 14–39.
55 *Game Protection Act 1866*, s13.
56 See Matthew K. Chew and Andrew L. Hamilton, "The Rise and Fall of Biotic Nativeness: A Historical Perspective," in *Fifty Years of Invasion Biology: The Legacy of Charles Elton*, ed. David Richardson (Oxford: Blackwell, 2011), 37.
57 *Birds Protection Act 1881*. See also Bonyhady, *Colonial*, 14–39; Dow, "'Sportsman's,'" 156–60; John M. Mackenzie, *The Empire of Nature* (Manchester: Manchester University Press, 1988), 25–53.
58 Dow, "'Sportsman's,'" 156–60.
59 *Birds Protection Act 1881*, s11; Jarman and Brock, "Evolving," 3; Gillian Hogendyk, "The Macquarie Marshes: An Ecological History," Institute of Public Affairs Occasional Paper, September 2007, 14.

60 Jarman and Brock, "Evolving," 3–6; Walker, "Fauna," 20.
61 Jarman and Brock, "Evolving," 6.
62 Walker, "Fauna," 19.
63 B. Nicholls (1925), quoted in Jarman and Brock, "Evolving," 7.
64 Jarman and Brock, "Evolving," 7. See also J. R. Kinghorn, "Bird Protection in Australia," *Emu* 28 (1929): 263–71.
65 Kinghorn, "Bird," 263.
66 Jarman and Brock, "Evolving," 6–7.
67 Ellis, "Ducks," 203; *Argus*, 1 July 1933, 27; *SMH*, 5 December 1936, 14.
68 *SMH*, 7 December 1937, 12; *Canberra Times*, 18 December 1937, 3; *Argus*, 8 December 1937, 17.
69 *Canberra Times*, 18 December 1937, 3.
70 *Argus*, 8 November 1938, 3. See also *SMH*, 29 November 1938, 7.
71 *SMH*, 14 November 1938, 3; *SMH*, 29 November 1938, 7; *SMH*, 9 December 1938, 3; *SMH*, 14 December 1938, 12; *SMH*, 22 December 1938, 4.
72 *SMH*, 9 December 1938, 3.
73 *SMH*, 29 November 1938, 7.
74 Jarman and Brock, "Evolving," 5; Pat Hutchings, "Foundations of Australian Science, Sydney's Natural History Legacy, and the Place of the Australian Museum," in *The Natural History of Sydney*, ed. Daniel Lunney, Pat Hutchings, and Dieter Hochuli (Mosman, NSW: Royal Zoological Society of NSW, 2012), 79.
75 *Argus*, 8 December 1937 17; *SMH*, 14 December 1938, 12.
76 *SMH*, 14 December 1938, 12.
77 *The Argus*, 8 December 1937, 17; *SMH*, 14 December 1938, 12.
78 Ellis, "Ducks," 200.
79 See also editor's note in Ellis, "Ducks," 202.
80 *Townsville Bulletin*, 26 October 1942, 2; *Camperdown Chronicle*, 13 March 1945, 4.
81 O'Gorman, "Growing," 107–8.
82 O'Gorman, *Flood*, 136–37.
83 Kabaila, *Wiradjuri*, 133 and 134.
84 *Canberra Times*, 3 February 1949, 5; *SMH*, 22 February 1952, 5.
85 For more on the 1951–52 season, see Frith, "Wild," 33; *SMH*, 26 September 1951, 2; *SMH*, 19 September 1952, 2; *The Mail* (Adelaide), 16 February 1952, 6; *Canberra Times*, 28 February 1952, 6; *SMH*, 1 September 1952, 4; Declarations of Special Open Seasons in NSW in 1950s, "NPWS Wildlife Files," 1961–88, W181, 12/12165 (B) (hereafter "NPWS"), SRNSW.
86 Frith, "Wild," 32; letter from D. V. Walters, Secretary Irrigation Research and Extension Committee, to Secretary, CSIRO, 22 February 1952, "Rice damage, goose, wild duck control in Australian rice fields," C2/8/2. A9778 (hereafter "Rice damage"), NAA.
87 Certificate of nomination to the Australian Academy of Science, 23 July 1971, H. J. Frith Files, Basser Library, Australian Academy of Science;

C. H. Tyndale-Biscoe, J. H. Calaby, and S. J. J. F. Davies, "Harold James Frith, 1921–1982," *Historical Records of Australian Science* 10, no. 3 (June 1995): 252.

88 Tyndale-Biscoe, Calaby, and Davies, "Harold James Frith," 247–49; Libby Robin, "Frith, Harold James (Harry) (1921–1982)," in *Australian Dictionary of Biography*, National Centre of Biography, Australian National University, http://adb.anu.edu.au/biography/frith-harold-james-harry-12517/text22523.

89 Letter from D. V. Walters, Secretary Irrigation Research and Extension Committee, to Secretary, CSIRO, 22 February 1952, "Rice damage," NAA.

90 Frith, "Wild," 33, 38, 40, 42, 43, 44, 45; memorandum from F. N. Ratcliffe to Secretary, CSIRO, 28 July 1952, "Rice damage," NAA; "NPWS," SRNSW.

91 Frith, "Wild," 33.

92 Frith, "Wild," 32–49. See also *Canberra Times*, 28 February 1952, 6; "NPWS," SRNSW.

93 Frith, "Wild," 36.

94 Frith, "Wild," 48–49.

95 Tyndale-Biscoe, Calaby, and Davies, "Harold James Frith," 252; *SMH*, 12 August 1939, 12.

96 Frith, "Wild," 49.

97 John Warhurst, "Ratcliffe, Francis Noble (1904–1970)," in *Australian Dictionary of Biography*, National Centre of Biography, Australian National University, 2002, http://adb.anu.edu.au/biography/ratcliffe-francis-noble-11490/text20491; Tyndale-Biscoe, Calaby, and Davies, "Harold James Frith," 250; Libby Robin, *Defending the Little Desert: The Rise of Ecological Consciousness in Australia* (Carlton South, Vic.: Melbourne University Press, 1998), 134–36; Libby Robin, *The Flight of the Emu: A Hundred Years of Australian Ornithology 1901–2001* (Carlton, Vic.: Melbourne University Press, 2001), 181–83; Robin, "Ecology," 69; Dunlap, *Nature*, 250–51; Martin Mulligan and Stuart Hill, *Ecological Pioneers: A Social History of Australian Thought and Action* (Cambridge: Cambridge University Press, 2001), 182–83.

98 Robin, "Ecology," 70. See also Warhurst, "Ratcliffe."

99 Sinclair, *Murray*, 175–76; B. Winterhalder, "Concepts in Historical Ecology: The View from Evolutionary Ecology," in *Historical Ecology: Cultural Knowledge and Changing Landscapes*, ed. C. Crumly (Santa Fe: School of American Research Press, 1993), 18–20; F. Golley, *A History of the Ecosystem Concept in Ecology: More Than the Sum of the Parts* (New Haven, CT: Yale University Press, 1993), 2.

100 Libby Robin, *How a Continent Created a Nation* (Sydney: University of New South Wales Press, 2007), 159.

101 Robin, *Continent*, 161–63; Robin, "Ecology," 70–71; Dunlap, *Nature*, 249–62; Mulligan and Hill, *Ecological*, 182–83.

102 Frith, "Wild," 33.

103 H. J. Frith, "Breeding Movements of Wild Ducks in Inland New South Wales," *Wildlife Research* 2, no. 1 (1957): 19–31.

104 See also H. J. Frith, *Waterfowl in Australia* (Sydney: Angus and Robertson, 1967); H. J. Frith, *Wildlife Conservation* (Sydney: Angus and Robertson, 1973).
105 H. J. Frith, unpublished book manuscript, H. J. Frith Files, Australian Academy of Science.
106 Jarman and Brock, "Evolving," 7.
107 Jarman and Brock, "Evolving," 8–9; Stubbs, "'Useless,'" 46.
108 Stubbs, "'Useless,'" 46; Jarman and Brock, "Evolving," 8.
109 Jarman and Brock, "Evolving," 8–9.
110 Kingsford, Webb, and Fullagar, "Scientific Panel," 12. See also essays in B. Dickson, J. Hutton, and B. Adams, eds., *Recreational Hunting, Conservation and Rural Livelihoods: Science and Practice* (West Sussex: Wiley-Blackwell, 2009).
111 *Fauna Protection Act 1948*, First Schedule. For open seasons, see, for example, "Duck Shooting NSW Rice Fields," 1957–78, VPRS 11559/P0001/300, PROV; "NPWS," SRNSW; Curtin and Kingsford, "Analysis," 24.
112 Curtin and Kingsford, "Ducks," 16–17.
113 Jarman and Brock, "Evolving," 11–12; "The Ramsar Convention on Wetlands," Ramsar Convention, www.ramsar.org.
114 Interview with Jamie Pittock and Cath Webb, November 2018; "Ramsar Convention."
115 See, for example, Edward O Wilson, "Foreword," in Richard T. T. Forman, *Land Mosaics: The Ecology of Landscapes and Regions* (Cambridge: Cambridge University Press, 1995), xiii–xiv.
116 Interview with anonymous farmer 2, May 2012.
117 Interview with Matthew Herring, August 2016.
118 "Bitterns Project," BirdLife Australia, https://birdlife.org.au/projects/bittern-project; "Bitterns in Rice," https://www.bitternsinrice.com.au/; interview with Mathew Herring, August 2016.
119 "Rice Farms Promote Biodiversity," Ricegrowers' Association of Australia, http://www.aboutrice.com/facts/fact06.html.
120 Curtin and Kingsford, "Ducks," 33.
121 For example, Coaral Rawton, "Making Your Dam 'Wildlife Friendly,'" Land for Wildlife Note No. 2, Land for Wildlife: Voluntary Wildlife Conservation, April 1999.

CHAPTER FIVE. ENCLOSING: PELICANS, PROTECTED AREAS, AND PRIVATE PROPERTY

This chapter has been adapted from "The Pelican Slaughter of 1911: A History of Competing Values, Killing and Private Property from the Coorong, South Australia," *Geographical Research* 54, no. 3 (2016): 285–300 (Wiley Publishers, © 2016, used by permission).

1 Heather Moorcroft and Michael Adams, "Emerging Geographies of Conservation and Indigenous Land in Australia," *Australian Geographer* 45, no. 4

(2014): 485–86; Heather Moorcroft, "Paradigms, Paradoxes and a Propitious Niche: Conservation and Indigenous Social Justice Policy in Australia," *Local Environment* 21, no. 5 (2016): 592; Lorena Pasquini, James A. Fitzsimons, Stuart Cowell, Katrina Brandon, and Geoff Wescott, "The Establishment of Large Private Nature Reserves by Conservation NGOs: Key Factors for Successful Implementation," *Oryx* 45, no. 3 (2011): 373–80; Ana Isla, "Conservation as Enclosure: An Ecofeminist Perspective on Sustainable Development and Biopiracy in Costa Rica," *Capitalism Nature Socialism* 16, no. 3 (2005): 49–61.

2 Heather Goodall, "Exclusion and Re-emplacement: Tensions around Protected Areas in Australia and Southeast Asia," *Conservation and Society* 4, no. 3 (2006): 385; Moorcroft, "Paradigms," 593.

3 Moorcroft and Adams, "Emerging," 488; Goodall, "Exclusion," 387; Robert Poirier and David Ostergren, "Evicting People from Nature: Indigenous Land Rights and National Parks in Australia, Russia, and the United States," *Natural Resources Journal* 42, no. 2 (2002): 334; Paige West, James Igoe, and Dan Brockington, "Parks and Peoples: The Social Impact of Protected Areas," *Annual Review of Anthropology* 35 (2006): 258.

4 Goodall, "Exclusion," 387–88; Poirier and Ostergren, "Evicting," 334.

5 Joan Schwartz and Terry Cook, "Archives, Records, and Power: The Making of Modern Memory," *Archival Science* 2, nos. 1–2 (2002): 1–19.

6 Bill Jeffery, "Cultural Contact along the Coorong in South Australia," *AIMA Bulletin* 25 (2001): 30; J. Kluske, "Coorong Park Notes," South Australian National Parks and Wildlife Service, 1991, Netley, South Australia, 124.

7 Kluske, "Coorong," 4; Jeffery, "Cultural," 30; "Draft Management Plan: Coorong National Park and Game Reserve," Department of Environment and Planning, Adelaide, 1984, 4, 9; Penny Rudduck, "European Heritage of the Coorong: A General Survey of the Sites of Early European Heritage of the Area Now Comprising the Coorong National Park and Coorong Game Reserve," National Parks and Wildlife Service, Government of South Australia, May 1982, 79.

8 F. R. H. Chapman, "The Pelican in South Australia with Special Reference to the Coorong Islands," *South Australian Ornithologist* 24 (1963): 6–14; Kluske, "Coorong," 79.

9 Julian Reid, "Australian Pelican: Flexible Responses to Uncertainty," in *Boom and Bust: Bird Stories for a Dry Country*, ed. Libby Robin, Robert Heinsohn, and Leo Joseph (Collingwood, Vic.: CSIRO, 2009), 96–97.

10 Reid, "Australian," 106.

11 Philip A. Clarke, "Birds as Totemic Beings and Creators in the Lower Murray, South Australia," *Journal of Ethnobiology* 36, no. 2 (2016): 282–87.

12 Reid, "Australian," 98 and 100–115.

13 "Coorong Game Reserve, Princes Hwy, Salt Creek via Meningie, SA, Australia," Australian Heritage Database, Department of the Environment, Australian Government, http://www.environment.gov.au/cgi-bin/ahdb/search.pl

?mode=place_detail;place_id=7909; Kluske, "Coorong," 81; Chapman, "Pelican," 8.
14 Today these are named Teal Island, North Pelican Island, Halfway Island, Pelican Island, Seagull Island, and Mellor Island. Kluske, "Coorong," 81.
15 Kluske, "Coorong," 81; Chapman, "Pelican," 8.
16 *DH*, 10 February 1911, 3; S. A. White, "Destruction of Pelicans," *Emu* 10, no. 5 (1911): 344. See also Chapman, "Pelican," 9.
17 Chapman, "Pelican," 8–9.
18 Irene Watson, "Buried Alive," *Law and Critique* 13, no. 3 (2002): 253–69; Mary Graham, "Some Thoughts about the Philosophical Underpinnings of Aboriginal Worldviews," *Worldviews: Environment, Culture, Religion* 3 (1999): 105–18.
19 P. A. Clarke, "Contact, Conflict and Regeneration: Aboriginal Cultural Geography of the Lower Murray" (PhD diss., University of Adelaide, South Australia, 1994), 80 and 225–27; Jeffery, "Cultural," 29–37; Kluske, "Coorong," 24 and 377; "Draft Management Plan," 24 and 53.
20 Jeffery, "Cultural," 30.
21 Clarke, "Contact," 53–54, 225–27, 42–62.
22 D. C. Paton, *At the End of the River: The Coorong and Lower Lakes* (Hindmarsh, South Australia: ATF Press, 2010), 153; A. M. Olsen, "The Coorong—a Multi-Species Fishery: Part 1—History and Development," Department of Fisheries, Adelaide, 1991, 5–13.
23 Olsen, "Coorong," 1 and 5–13.
24 Chapman, "Pelican," 11.
25 Chapman, "Pelican," 11.
26 Chapman, "Pelican," 12.
27 *DH*, 9 February 1911, 6.
28 *Fisheries Act Amendment 1909*, s5.
29 *Fisheries Act 1904*, s7; *Fisheries Act Amendment 1909*, s5; *The Advertiser*, 4 February 1910, 6.
30 *DH*, 10 February 1911, 3; *DH*, 10 February 1911, 4; *Fisheries Act Amendment 1909*, s5.
31 *Bird Protection Act 1900*, Schedule 2. See also Joanna Sassoon, "'The Common Cormorant or Shag Lays Eggs inside a Paper Bag': A Cultural Ecology of Fish-Eating Birds in Western Australia," *Environment and History* 9, no. 1 (2003): 36.
32 *Bird Protection Act 1900*; *DH*, 10 February 1911, 3; *DH*, 10 February 1911, 4.
33 *South Australian Government Gazette*, 24 December 1908, 1255.
34 *Bird Protection Act 1900*, s3 and s4.
35 *The Register*, 13 May 1910, 4; *Renmark Pioneer*, 20 May 1910, 8; *The Register*, 9 August 1910, 2.
36 *DH*, 10 February 1911, 4; *The Advertiser*, 4 February 1910, 6.
37 *DH*, 10 February 1911, 3.
38 *Fisheries Act Amendment 1909*, s5.

39 Pre-Decimal Currency Converter, Reserve Bank of Australia, http://www.rba.gov.au/calculator/annualPreDecimal.html.
40 R. Crompton, "History of Ornithology in South Australia," *South Australian Ornithologist* 1, no. 1 (1914): 8–9 and 11.
41 R. W. Linn, "White, Samuel Albert (1870–1954)," *Australian Dictionary of Biography*, National Centre of Biography, Australian National University, http://adb.anu.edu.au/biography/white-samuel-albert-9079/text16007 (published in hardcopy 1990); "Protection of Pelicans in South Australia," *Emu* 11, no. 1 (1911): 45; *The Advertiser*, 9 February 1911, 11; "Mellor, John White (1868–1931)," in *Encyclopaedia of Australian Science*, http://www.eoas.info/biogs/P003173b.htm; Robin, *Flight*, 36; *The Advertiser*, 3 June 1910, 5; *The Advertiser*, 2 June 1910, 11; *The Advertiser*, 9 February 1911, 11.
42 *DH*, 10 February 1911, 3.
43 *DH*, 10 May 1911, 6; *The Advertiser*, 2 June 1910, 11. See also White, "Destruction."
44 *DH*, 10 February 1911, 3.
45 K. Thomas, *Man and the Natural World: Changing Attitudes in England 1500–1800* (London: Penguin, 1991); Sassoon, "Common," 31–52.
46 *Adelaide Evening Journal*, 9 February 1911, quoted in White, "Destruction." See also Robin W. Doughty, *Feather Fashions and Bird Preservation: A Study in Nature Protection* (Berkeley: University of California Press, 1975), 112.
47 *DH*, 10 May 1911, 4; *DH*, 10 May 1911, 6; Libby Robin, "Emu: National Symol and Ecological Limits," in *Boom and Bust: Bird Stories for a Dry Country*, ed. Libby Robin, Robert Heinsohn, and Leo Joseph (Collingwood, Vic.: CSIRO, 2009), 250.
48 *The Advertiser*, 6 March 1911, 11.
49 *The Register*, 11 February 1911, 12; *The Register*, 10 February 1911, 4; *DH*, 10 May 1911, 6; *Argus*, 10 February 1911, 8; *Barrier Miner*, 14 February 1911, 3; *Cairns Post*, 11 February 1911, 3.
50 *DH*, 10 May 1911, 4.
51 *DH*, 10 February 1911, 3.
52 *DH*, 10 February 1911, 3; *The Register*, 13 February 1911, 6.
53 *Northern Star*, 14 February 1911, 6; *Cairns Post*, 11 February 1911, 3; *DH*, 10 May 1911, 6; *The Register*, 10 February 1911, 4; *The Advertiser*, 6 March 1911, 11.
54 *DH*, 15 February 1911, 3.
55 *DH*, 15 February 1911, 3.
56 Daniel Sarewitz, "How Science Makes Environmental Controversies Worse," *Environmental Science and Policy* 7, no. 5 (2004): 385–403.
57 Emily O'Gorman, "Remaking Wetlands: Rice Fields and Ducks in the Murrumbidgee River Region, NSW," in *Rethinking Invasion Ecologies from the Environmental Humanities*, ed. Jodi Frawley and Iain McCalman (London: Routledge, 2014), 215–38.
58 Dunlap, *Nature*, 52.
59 Dunlap, *Nature*, 1–4; Mark Cioc, *The Game of Conservation: International Treaties to Protect the World's Migratory Animals* (Athens: Ohio University Press, 2009).

60 Michael A. Osborne, "Acclimatizing the World: A History of the Paradigmatic Colonial Science," *Osiris* (2000): 135–51; Dunlap, *Nature*, 52–59.
61 Tim Bonyhady, *The Colonial Earth* (Melbourne: Melbourne University Publishing, 2003), 14–39; Dow, "'Sportsman's,'" 148; Walker, "Fauna," 18; Jarman and Brock, "Evolving," 2; Stubbs, "'Useless,'" 26–28; B. C. Newland, "From Game Laws to Fauna Protection Acts in South Australia: The Evolution of an Attitude," *South Australian Ornithologist* 23 (1961): 53.
62 *Game Act 1874*.
63 Jarman and Brock, "Evolving," 6.
64 Newland, "Game," 54–55.
65 Newland, "Game," 53–55; Steven White, "British Colonialism, Australian Nationalism and the Law: Hierarchies of Wild Animal Protection," *Monash University Law Review* 39, no. 2 (2013): 452–72; Penny Olsen, *Upside Down World: Early European Impressions of Australia's Curious Animals* (Canberra: National Library Australia, 2010), 8.
66 *DH*, 10 May 1911, 6; Newland, "Game," 55. For example, *South Australian Chronicle and Weekly Mail*, 3 April 1880, 2 and 3.
67 White, "British," 470.
68 Newland, "Game," 53.
69 Newland, "Game," 52–54; Paton, *End*, 110; *Game Act 1874*, s12; *Game Act 1886*, s12; *Bird Protection Act 1900*, s4; *Animals Protection Act 1912*, s18; *Animals and Birds Protection Act 1919*, s20.
70 Philip A. Clarke, "Aboriginal Foraging Practices and Crafts involving Birds in the Post-European Period of the Lower Murray, South Australia," *Transactions of the Royal Society of South Australia* 142, no. 1 (2018): 11–15.
71 Watson, "Buried," 261–62; Lauren Butterly, "'For the Reasons Given in Akiba . . .': Karpany V. Dietman [2013] Hca 47," *Indigenous Law Bulletin* 8, no. 10 (2014): 23–26.
72 Alex Cuthbert Castles, *An Australian Legal History* (Sydney: Law Book Company, 1982), 524–25; Watson, "Buried," 262; Heather Douglas and Mark Finnane, *Indigenous Crime and Settler Law: White Sovereignty after Empire* (London: Palgrave Macmillan, 2012).
73 Henry Reynolds, *Aboriginal Sovereignty: Reflections on Race, State and Nation* (Sydney: Allen and Unwin, 1996), 121; Watson, "Buried," 262.
74 White, "British," 466.
75 *Vermin Destruction Act 1882*, s2.
76 Olsen, *Upside Down*, 10–11.
77 Deborah Rose, "What If the Angel of History Were a Dog?," *Cultural Studies Review* 12, no. 1 (2013): 67–78. See also White, "British," 462.
78 Jarman and Brock, "Evolving," 3–6; Walker, "Fauna," 20.
79 White, "British," 467–68.
80 Doughty, *Feather*, esp. 30; Jennifer Price, *Flight Maps: Adventures with Nature in Modern America* (New York: Basic Books, 2000), 56–109; Dunlap, *Nature*, 124.

81 Doughty, *Feather*, 154; Robin, *Continent*, 25.
82 Jarman and Brock, "Evolving," 7. Cioc, *Game*, 64; Dunlap, "Ecology," 77–81.
83 Crompton, "History," 9.
84 Crompton, "History," 9. During the nineteenth and early twentieth centuries many birds were killed and eggs preserved for museum collections around Australia. Robin, *Continent*, 78–79; Dunlap, *Nature*, 92–95.
85 Crompton, "History," 9.
86 Newland, "Game," 58. On humanitarian motives for protection and anti-cruelty laws, see White, "British."
87 "Protection of Pelicans," 45.
88 Paton, *End*, 109–13; *A Trip to the Coorong, Christmas 1902* and *Off the Chain: Holiday Experiences on Lake Coorong February 190*, Local History Archive, Alexandrina Library.
89 "Protection of Pelicans," 45.
90 The islands included in the lease lay "between Wood's Well and Salt Creek." C. Barrett, "The Coorong Islands," *Emu* 11 (1911): 127–28; SOA, "Re Islands in Coorong," 1911, GRG35/1/00000/353, File 1215/1911, SOA, SRSA. See also Chapman, "Pelican," 13.
91 Cioc, *Game*, 58–103.
92 SOA, SRSA.
93 SOA, SRSA.
94 Sassoon, "'Common."
95 P. A. Clarke, "Twentieth-Century Aboriginal Harvesting Practices in the Rural Landscape of the Lower Murray," *South Australia. Records of the South Australian Museum* 36, (2003): 83–107.
96 Barrett, "The Coorong Islands," 128.
97 There is some evidence to suggest that the initial bird protection district in the region, established in 1908, was partly intended by local government officials—ironically led by a Fisheries Inspector—to stop "whites and natives" from collecting swan and pelican eggs, in addition to preventing shooters from overhunting birds. *The Advertiser*, 12 December 1908, 11; *Observer*, 19 December 1908, 36; *Chronicle*, 26 December 1908, 39; *The Advertiser*, 6 January 1909, 8; *The Register*, 6 January 1909, 3; *The Register*, 10 November 1906, 6.
98 Clarke, "Aboriginal Foraging," 14.
99 SOA, SRSA.
100 Clarke, "Contact," 80–82; Marcia Langton, *Well, I Heard It on the Radio and I Saw It on the Television*, vol. 9 (Sydney: Australian Film Commission, 1993), 31–43; Mitchell Rolls, "The Meaninglessness of Aboriginal Cultures," *Balayi* 2, no. 1 (2001): 9–10.
101 See Clarke, "Contact"; Rolls, "Meaninglessness."
102 SOA, SRSA.
103 SOA, SRSA.
104 Many Aboriginal people and scholars have problematized these ideologies, embedded in language, within the postcolonial movements of the last few

decades. Notions of "blood purity" have also been examined by many Indigenous scholars as a problematic notion of identity that has carried forward discourses of race and assimilation. See discussion in Rolls, "Meaninglessness."

105 *DH*, 15 February 1911, 3.
106 *Animals Protection Act* 1912, s18; *Animals and Birds Protection Act* 1919, s21.
107 Clarke, "Contact," 332; Clarke, "Aboriginal Foraging," 13–14.
108 Clarke, "Contact," 337.
109 M. J. Adams, V. Cavanagh, and B. Edmunds, "Bush Lemons and Beach Hauling: Evolving Traditions and New Thinking for Protected Areas Management and Aboriginal People in New South Wales," in *Protecting Country: Indigenous Governance and Management of Protected Areas*, ed. D. Smyth and G. R. Ward (Canberra: AIATSIS, 2008), 31–45.
110 SOA, SRSA.
111 H. T. Condon, "Young Pelicans," *Emu* 41, no. 1 (1941): 92. See also Chapman, "Pelican," 11–13.
112 Chapman, "Pelican," 8.
113 Chapman, "Pelican," 13; Richard T. Kingsford, Keith F. Walker, Rebecca E. Lester, William J. Young, Peter G. Fairweather, Jesmond Sammut, and Michael C. Geddes, "A Ramsar Wetland in Crisis—the Coorong, Lower Lakes and Murray Mouth, Australia," *Marine and Freshwater Research* 62, no. 3 (2011): 255–65; R. T. Kingsford and R. F. Thomas. "Destruction of Wetlands and Waterbird Populations by Dams and Irrigation on the Murrumbidgee River in Arid Australia," *Environmental Management* 34, no. 3 (2004): 383–96.
114 Reid, "Australian," 105.
115 Jenny Lau, "Birds of the Murray-Darling Basin: Overview," in *Birds of the Murray-Darling Basin*, ed. Richard Kingsford, Jenny Lau, and James Connor, BirdLife Australia Conservation Statement No. 16, May 2014, 3.
116 Rose, "What If"; Deborah Bird Rose, *Wild Dog Dreaming: Love and Extinction* (Charlottesville: University of Virginia Press, 2011); Freya Mathews, "Against Kangaroo Harvesting," *Journal of Bioethical Inquiry* 10, no. 2 (2013): 263–65.

CHAPTER SIX. MIGRATING: WETLANDS, TRANSCONTINENTAL BIRD MOVEMENTS, AND GLOBAL ENVIRONMENTAL CRISIS

1 Shultz and Phillips, "Fivebough and Tuckerbil."
2 Geoffrey C. Bowker and Susan Leigh Star, *Sorting Things Out: Classification and Its Consequences* (Cambridge: MIT Press, 2000), esp. 190.
3 Bruce Braun, *The Intemperate Rainforest: Nature, Culture, and Power on Canada's West Coast* (Minneapolis: University of Minnesota Press, 2002), 3.
4 While Ngrams generated by Google are far from complete records (they are based only on digitized books), this tool gives an indication of the almost meteoric rise of the term "wetlands" from the 1960s: https://books.google.com/ngrams/graph?content=wetland%2C+wetlands&year_start=1800&year_end

=2000&corpus=15&smoothing=3&share=&direct_url
=t1%3B%2Cwetland%3B%2Cc0%3B.t1%3B%2Cwetlands%3B%2Cc0.

5 Paul Warde, Libby Robin, and Sverker Sörlin, *The Environment: A History of the Idea* (Baltimore: Johns Hopkins University Press, 2018), 1–3; Alessandro Antonello, *The Greening of Antarctica: Assembling an International Environment* (Oxford: Oxford University Press, 2019).

6 Braun, *Intemperate*, 3.

7 See Cioc, *Game*, 58–101; Wilson, *Seeking*; Libby Robin, "Birds and Environmental Management in Australia 1901–2001," *Australian Journal of Environmental Management* 8, no. 2 (2001):105–13; Nancy Jacobs, *Birders of Africa: History of a Network* (New Haven, CT: Yale University Press, 2016); Kirsten Greer, "Geopolitics and the Avian Imperial Archive: The Zoogeography of Region-Making in the Nineteenth-Century British Mediterranean," *Annals of the Association of American Geographers* 103, no. 6 (2013): 1317–31; Robert Wilson, "Mobile Bodies: Animal Migration in North American History," *Geoforum* 25 (2015): 465–72.

8 "UNESCO—Convention of Wetlands of International Importance—Waterfowl Habitat," A1838, 862/11/9 (hereafter "UNESCO"), Part 1, NAA; Geoffrey Vernon Townsend Matthews, "The Ramsar Convention on Wetlands: Its History and Development" (Gland, Switzerland: Ramsar Convention Bureau, 1993), 4.

9 "Environment—Relations with Other Countries," A1838, 703/3 (hereafter "Environment"), Parts 1–4, NAA.

10 Mulligan and Hill, *Ecological*; Drew Hutton and Libby Connors, *History of the Australian Environment Movement* (Cambridge: Cambridge University Press, 1999).

11 "UNESCO," Part 1, NAA.

12 House of Representatives Select Committee, *Wildlife Conservation: Report from the House of Representatives Select Committee* (Canberra: Government Printer of Australia, 1972), 11.

13 Committee, *Wildlife*, 13.

14 Committee, *Wildlife*, 55–56.

15 H. Frith, F. Crome, and B. Brown, "Aspects of the Biology of the Japanese Snipe Gallinago hardwickii," *Australian Journal of Ecology* 2, no. 3 (1977): 341.

16 Frith, Crome, and Brown, "Aspects."

17 Committee, *Wildlife*, 56.

18 Committee, *Wildlife*, 26.

19 Matthews, "Ramsar," 5–13.

20 Robert Boardman, *International Organization and the Conservation of Nature* (London: Macmillan, 1981), 62.

21 Robert Boardman, *The International Politics of Bird Conservation: Biodiversity, Regionalism and Global Governance* (Cheltenham: Edward Elgar, 2006), 50.

22 "Ramsar Convention on Wetlands of International Importance Especially as Waterfowl Habitat" (Paris: UNESCO, 1994), original text 1971, amended 1982 and 1987. See "Ramsar Convention."

23 For example, Eben Kirskey, "Living with Parasites in Palo Verde National Park," *Environmental Humanities* 1, no. 1 (2012): 23–55.
24 For example, Xander Combrink, Jan L. Korrûbel, Robert Kyle, Ricky Taylor, and Perran Ross, "Evidence of a Declining Nile Crocodile (Crocodylus Niloticus) Population at Lake Sibaya, South Africa," *African Journal of Wildlife Research* 41, no. 2 (2011): 145–57; W. K. Saalfeld, R. Delaney, Y. Fukuda, and A. J. Fisher, "Management Program for the Saltwater Crocodile in the Northern Territory of Australia, 2014–2015" (Darwin: Northern Territory Department of Land Resource Management, 2014).
25 Boardman, *International Organization*; John McCormick, *Reclaiming Paradise: The Global Environmental Movement* (Bloomington: Indiana University Press, 1991). See also "Environment," Parts 1–4, NAA.
26 Michael E. Soulé, "What Is Conservation Biology?," *BioScience* 35, no. 11 (1985): 727–34; Libby Robin, "The Rise of the Idea of Biodiversity: Crises, Responses and Expertise," *Quaderni: Communication, Technologies, Pouvoir* 76 (2011): 25–37.
27 Matthews, "Ramsar"; Wilson, *Seeking*; Garone, *Fall*; Vileisis, *Discovering*. An early journal in wetlands ecology, *Wetlands* was first established in 1981. In the first issue of *Wetlands Ecology and Management*, published in 1989, the editor Rebecca Sharitz wrote: "The study of wetlands, or of wetland ecology, is developing as a distinct field of scientific interest." R. R. Sharitz, "Comments of the Editor," *Wetlands Ecology Management* 1, no. 1 (1989): 1.
28 A. C. Martin, N. Hotchkiss, F. M. Uhler, and W. S. Bourn, *Classification of Wetlands of the United States*, No. 20 (Washington, DC: US Fish and Wildlife Service, 1953); National Research Council, *Wetlands: Characteristics and Boundaries* (Washington, DC: National Academies Press, 1995).
29 Sylvia Mary Haslam, *Understanding Wetlands: Fen, Bog and Marsh* (Boca Raton, FL: CRC Press, 2004), 1–2; Martin et al., *Classification*; National Research Council, *Wetlands*.
30 "UNESCO," Part 1, NAA.
31 Richard T. Kingsford, Alberto Basset, and Leland Jackson, "Wetlands: Conservation's Poor Cousins," *Aquatic Conservation: Marine and Freshwater Ecosystems* 26, no. 5 (2016): 893.
32 Cioc, *Game*, 149–52.
33 Matthews, "Ramsar."
34 Marcia Langton, "What Do We Mean By Wilderness? Wilderness and Terra Nullius in Australian Art [address to the Sydney Institute on 12 October 1995]," *Sydney Papers* 8, no. 1 (1996): 10; Mulligan and Hill, *Ecological*, 154, 245, 285; Goodall, "Exclusion," 386.
35 Helen Ross, Chrissy Grant, Cathy J. Robinson, Arturo Izurieta, Dermot Smyth, and Phil Rist, "Co-management and Indigenous Protected Areas in Australia: Achievements and Ways Forward," *Australasian Journal of Environmental Management* 16, no. 4 (2009): 244.
36 Goodall, "Exclusion," 384–87.

37 "Environment," Parts 1 and 2, NAA.
38 Anne Twomey, "Federal Parliament's Changing Role in Treaty Making and External Affairs," Parliament of Australia, https://www.aph.gov.au/About_Parliament/Parliamentary_Departments/Parliamentary_Library/pubs/rp/rp9900/2000RP15.
39 "Environment," Parts 1 and 2, NAA.
40 "UNESCO," Part 1, NAA; "Convention on Wetlands of International Importance [Box 134]," A432, 73/3337, 1973–75 (hereafter "Convention"), NAA; "Environment," Parts 1 and 2, NAA; Twomey, "Federal"
41 "UNESCO," Part 1, NAA; "Convention," NAA. Australian National Parks and Wildlife Service was renamed the Australian Conservation Agency in the 1990s. "Australian National Parks and Wildlife Service," *Encyclopedia of Australian Science*, http://www.eoas.info/biogs/A001846b.htm
42 "Convention," NAA.
43 Libby Robin, "Frith, Harold James (Harry) (1921–1982)," in *Australian Dictionary of Biography*, National Centre of Biography, Australian National University, http://adb.anu.edu.au/biography/frith-harold-james-harry-12517/text22523 (published first in hardcopy 2007).
44 Robin, "Frith"; Emily O'Gorman, "Remaking Wetlands: Rice Fields and Ducks in the Murrumbidgee River Region, NSW," in *Rethinking Invasion Ecologies from the Environmental Humanities*, ed. Jodi Frawley and Iain McCalman (London: Routledge Environmental Humanities, 2014), 215–38; Frith, Crome, and Brown, "Aspects," 341–68.
45 "Convention," NAA.
46 "Convention," NAA.
47 "Convention," NAA.
48 "Environment," Part 1, NAA. The Japan agreement did not enter into force until 1981. "Agreement between the Government of Australia and the Government of Japan for the Protection of Migratory Birds in Danger of Extinction and their Environment" (Japan-Australia Migratory Bird Agreement, JAMBA), Australian Treaty Series 1981, No. 6.
49 "Environment," Part 1, NAA.
50 "Environment," Parts 1–4, NAA.
51 Libby Robin, "Migrants and Nomads: Seasoning Zoological Knowledge in Australia," in *A Change in the Weather: Climate and Culture in Australia*, ed. Tim Sherratt, Tom Griffiths, and Libby Robin (Canberra: National Library of Australia, 2005), 42–53. See also Jacobs, *Africa*, 92–120.
52 *The Australasian*, 29 July 1893, 32.
53 Jacobs, *Africa*, 92–120. See also Greer, "Geopolitics."
54 *The Australasian*, 12 September 1891, 39. Attributed to "The Field Naturalist." Likely written by Campbell due to his frequent columns for the newspaper, mention of services of "Mr. Owston," and similarity to later texts by Campbell. Henry Seebohm also described the possible breeding place and migration of this snipe in 1887. Henry Seebohm, *The Geographical Distribution of the Family*

Charadriidae, or the Plovers, Sandpipers, Snipes, and Their Allies (London: Henry Sotheran, 1887), 473–74; Henry Seebohm, *Birds of the Japanese Empire* (London: R. H. Porter, 1890), 342.

55 A. J. Campbell, *Nests and Eggs of Australian Birds (Part III)* (Sheffield: Pawson and Brailsford, 1901), 822–26. See also *The Australasian*, 29 July 1893, 32.
56 "Indigenous Weather Knowledge," Australian Bureau of Meteorology, http://www.bom.gov.au/iwk/culture.shtml.
57 Philip A. Clarke, "The Ngarrindjeri Nomenclature of Birds in the Lower Murray River Region, South Australia," *Transactions of the Royal Society of South Australia* 143, no. 1 (2019): 122–23.
58 Philip A. Clarke, "Birds as Totemic Beings and Creators in the Lower Murray, South Australia," *Journal of Ethnobiology* 36, no. 2 (2016): 282.
59 Rose, *Reports*; Clarke, "Totemic Beings."
60 Robin, "Migrants."
61 For example, R. Hall, "The Eastern Paleaerctic," *Emu* 19, no. 2 (1919): 82–98.
62 For example, Sergius Buturlin, "Australian Birds in Siberia," *Emu* 11, no. 2 (1911): 95.
63 Robin, "Migrants," 43–48.
64 R. W. Legge, "Australian Snipe," *Emu* 31, no. 4 (1931): 308; Frith, Crome, and Brown, "Aspects."
65 "Gallinago hardwickii—Latham's Snipe, Japanese Snipe," Department of Environment, Australian Government, http://www.environment.gov.au/cgi-bin/sprat/public/publicspecies.pl?taxon_id=863.
66 Boardman, *International Politics*, 155; K. Dorsey, *The Dawn of Conservation Diplomacy: US-Canadian Wildlife Protection Treaties in the Progressive Era* (Seattle: University of Washington Press, 2009).
67 Robin, "Migrants," 46–53.
68 Robin, "Migrants," 46–53.
69 Robin, *Flight*, 160.
70 See also Peter Coates, *American Perceptions of Immigrant and Invasive Species: Strangers on the Land* (Berkeley: University of California Press, 2007).
71 *World's News*, 10 July 1954, 14.
72 *World's News*, 31 October 1953, 15.
73 Robin, *Flight*, 246–47; H. Elliott McClure, "Migratory Animal Pathological Survey: Progress Report 1967" (Washington, DC: US Army, 1968); Michael Lewis, "Scientists or Spies? Ecology in a Climate of Cold War Suspicion," *Economic and Political Weekly*, 15 June 2005, 2326–30; Michael Lewis, *Inventing Global Ecology: Tracking the Biodiversity Ideal in India, 1947–1997* (Athens: Ohio University Press, 2004); Robert Wilson, "Directing the Flow: Migratory Waterfowl, Scale, and Mobility in Western North America," *Environmental History* 7, no. 2 (2002): 247–66.
74 "Environment," Part 1, NAA.
75 "Environment," Parts 2 and 3, NAA, "Agreement between the Government of Australia and the Government of Japan."

76 "Environment," Part 1, NAA.
77 "Environment," Part 1, NAA.
78 "Environment," Part 1, NAA; Rica Erickson, "Serventy, Dominic Louis (1904–1988)," in *Australian Dictionary of Biography*, National Centre of Biography, Australian National University, http://adb.anu.edu.au/biography/serventy-dominic-louis-15496/text26711 (published first in hardcopy 2012).
79 "Division of Wildlife Research. Conservation. Convention with Japan on Migratory Birds," A9697, C4/81, NAA.
80 "Environment," Part 3, NAA.
81 G. N. Goodrick, *A Survey of Wetlands of Coastal New South Wales*, CSIRO Division of Wildlife Research. Technical Memorandum No. 5 (Canberra: CSIRO, 1970).
82 "UNESCO," Part 1, NAA; Kees Paijmans, "Feasibility Report on a National Wetland Survey," Division of Land Use Research, CSIRO, Technical Memorandum 78/6, May 1978.
83 "CSIRO—Division of Wildlife Research—Conservation—Wetlands Survey," A9697, C4/86 (hereafter "CSIRO"), NAA; see also Paijmans, "Feasibility."
84 "CSIRO," NAA.
85 "CSIRO," NAA.
86 "CSIRO," NAA.
87 James C. Scott, *Seeing Like a State: How Certain Schemes to Improve the Human Condition Have Failed* (New Haven, CT: Yale University Press, 1998), 11–52.
88 Paijmans, "Feasibility"; "CSIRO," NAA.
89 "CSIRO," NAA. See also Paijmans, "Feasibility."
90 "CSIRO," NAA.
91 "CSIRO," NAA. Land Use research completed a further feasibility study in 1978, endorsed by CONCOM.
92 Richard T. Kingsford, "Conservation of Floodplain Wetlands—Out of Sight, Out of Mind?," *Aquatic Conservation: Marine and Freshwater Ecosystems* 25, no. 6 (2015): 727–32; Kingsford, Basset, and Jackson, "Wetlands"; C. M. Finlayson, J. A. Davis, A. Gell, R. T. Kingsford, and K. A. Parton, "The Status of Wetlands and the Predicted Effects of Global Climate Change: The Situation in Australia," *Aquatic Sciences* 75, no. 1 (2013): 73–93; C. M. Finlayson, "Loss and Degradation of Australian Wetlands," paper for the LAW ASIA Conference "Environmental Law Issues in the Asia-Pacific Region," Darwin, Australia, 2000; Colin M. Finlayson and Naomi Rea, "Reasons for the Loss and Degradation of Australian Wetlands," *Wetlands Ecology and Management* 7, nos. 1–2 (1999): 3; C. M. Finlayson and D. S. Mitchell, "Australian Wetlands: The Monitoring Challenge," *Wetlands Ecology and Management* 7, nos. 1–2 (1999): 105–12.
93 G. Bino, R. T. Kingsford, and K. Brandis, "Australia's Wetlands—Learning from the Past to Manage for the Future," *Pacific Conservation Biology* 22, no. 2 (2016): 117.
94 Finlayson and Rea, "Reasons," 3.

95 Kingsford, Basset, and Jackson, "Wetlands," 900. See also Bino, Kingsford, and Brandis, "Australia's Wetlands"; Finlayson et al., "Status of Wetlands," 74.
96 Bino, Kingsford, and Brandis, "Australia's wetlands," 117. See also R. L. Pressey, "Some Problems with Wetland Evaluation," *Wetlands Australia* 5, no. 1 (2010): 42–51; R. L. Pressey and P. Adam, "A Review of Wetland Inventory and Classification in Australia," *Vegetatio* 118 (1991): 81–101.
97 Kingsford, "Conservation," 727.
98 Kingsford, "Conservation," 730.
99 "CSIRO," NAA.
100 Murray-Darling Basin Authority, "Birds of the Murray-Darling Basin: 2017 Basin Plan Evaluation," December 2017, 3.
101 Complied from information available from "Australia's Ramsar Sites," Australian Government Department of Environment and Energy, https://www.environment.gov.au/water/wetlands/australian-wetlands-database/australian-ramsar-wetlands.
102 Murray-Darling Basin Authority, "Birds of the Murray-Darling Basin," 4.
103 Boardman, *International Politics*, 157 and 164–65.
104 J. Pittock, M. Finlayson, A. Gardner, and C. McKay, "Changing Character: The Ramsar Convention on Wetlands and Climate Change in the Murray-Darling Basin, Australia," *Environmental and Planning Law Journal* 27 (2010): 417. The Commonwealth Government of Australia's obligations to the Ramsar Convention are largely supported by both the Water Act 2007 and the Environment Protection and Biodiversity Conservation Act 1999. See D. R. Rothwell, S. Kaye, A. Akhtarkhavari, and R. Davis, *International Law: Cases and Materials with Australian Perspectives* (Cambridge: Cambridge University Press, 2010), 590; C. A. Brebbia, *Sustainable Development and Planning Vi* (Ashurst, Southampton: WIT Press, 2013), 142; "Australia's Obligations under the Ramsar Convention: Legislative Support for Wetlands," Australian Government Department of Environment and Energy, http://www.environment.gov.au/water/wetlands/publications/australias-obligations-under-ramsar-convention-legislative-support-wetlands-fact-sheet.
105 Connell, *Water*, 178.
106 Pittock et al., "Changing," 405.
107 Pittock et al., "Changing," 425.
108 D. A. Scott and T. A. Jones, "Classification and Inventory of Wetlands: A Global Overview," *Vegetatio* 118, nos. 1–2 (1995): 4 and 6.
109 "Fivebough and Tuckerbil Swamps," NSW Department of Planning, Industry and Environment, http://www.environment.nsw.gov.au/topics/water/wetlands/internationally-significant-wetlands/fivebough-and-tuckerbil-swamps.
110 "Eastern Australian Waterbird Survey," Centre for Ecosystem Science, University of New South Wales, https://www.ecosystem.unsw.edu.au/content/rivers-and-wetlands/waterbirds/eastern-australian-waterbird-survey; R. T. Kingsford and J. L. Porter, "Monitoring Waterbird Populations with Aerial Surveys—What Have We Learnt?," *Wildlife Research* 36, no. 1 (2009): 29–40.

111 Kingsford, Basset, and Jackson, "Wetlands," 900.
112 Finlayson and Mitchell, "Australian," 108
113 See examples in Robin, "Birds," 108–9.
114 "Fivebough and Tuckerbil Swamps."
115 Robin, "Birds," 109.
116 Finlayson et al., "Status of Wetlands," 83.
117 Sandra McGregor, Violet Lawson, Peter Christophersen, Rod Kennett, James Boyden, Peter Bayliss, Adam Liedloff, Barbie McKaige, and Alan N. Andersen, "Indigenous Wetland Burning: Conserving Natural and Cultural Resources in Australia's World Heritage-listed Kakadu National Park," *Human Ecology* 38, no. 6 (2010): 721–29.
118 "Ngarrindjeri Regional Authority," Ngarrindjeri Regional Authority, https://www.ngarrindjeri.org.au/.
119 Weir, Crew, and Crew, "Wetland."

CHAPTER SEVEN. RIPPLING: CAPITALISM, SEALS, AND BASELINES

1 Jeffery, "Cultural," 29–30; Kluske, "Coorong," 14–15; "Draft Management Plan," 53.
2 Claire Maree Settre and Sarah Ann Wheeler, "A Century of Intervention in a Ramsar Wetland—the Case of the Coorong, Lower Lakes and Murray Mouth," *Australasian Journal of Environmental Management* 24, no. 2 (2017): 163–83; P. Gell and D. Haynes, "A Palaeoecological Assessment of Water Quality Change in the Coorong, South Australia," report for the Department of Water, Land and Biodiversity Conservation, University of Adelaide, Adelaide, 2005.
3 For example, Gell and Haynes, "Palaeoecological."
4 S. Hemming and D. Rigney, "Ngarrindjeri Futures: Negotiating a Future through Caring for Ruwe/Ruwar (Lands, Waters and All Living Things)," in *Innovation for 21st Century Conservation*, ed. P. Figgis, J. Fitzsimons and J. Irving (Sydney: Australian Committee for IUCN, 2012), 186–91; R. Hattam, D. Rigney, and S. Hemming, "Reconciliation? Culture, Nature and the Murray River," in *Fresh Water: New Perspectives on Water in Australia*, ed. E. Potter, A. Mackinnon, S. McKenzie, and J. McKay (Carlton, Vic.: Melbourne University Press, 2007), 105–22; S. Hemming and D. Rigney, "Restoring Murray Futures: Incorporating Indigenous Knowledge, Values and Interests into Environmental Water Planning in the Coorong and Lakes Alexandrina and Albert Ramsar Wetland," Goyder Institute for Water Research Technical Report Series No. 16/8, Adelaide, South Australia, 2016.
5 Working Group on Long-Nosed Fur Seals in the Coorong and Lower Lakes, "Communiqué no. 15," Department of Environment, Water and Natural Resources, Government of South Australia, February 2018.
6 P. D. Shaughnessy, S. D. Goldsworthy, and A. I. Mackay, "The Long-Nosed Fur Seal (Arctocephalus forsteri) in South Australia in 2013–14: Abundance, Status

and Trends," *Australian Journal of Zoology* 63, no. 2 (2015): 101–10; B. L. Chilvers and S. D. Goldsworthy, *Arctocephalus forsteri: The IUCN Red List of Threatened Species* (2015), e.T41664A45230026, http://dx.doi.org/10.2305/IUCN.UK.2015-2.RLTS.T41664A45230026.en.

7 "Community Questions on Long-Nosed Fur Seals," South Australia Department of Environment and Water, https://www.environment.sa.gov.au/managing-natural-resources/plants-and-animals/Living_with_wildlife/seals/community-questions-long-nosed-fur-seals.

8 Interview with Garry Hera-Singh, March 2012.

9 Peter S. Alagona, John Sandlos, and Yolanda F. Wiersma, "Past Imperfect: Using Historical Ecology and Baseline Data for Conservation and Restoration Projects in North America," *Environmental Philosophy* 9, no. 1 (2012): 50.

10 Haraway, *Species*, 45–65; M. Barua, "Lively Commodities and Encounter Value," *Environment and Planning D: Society and Space* 34, no. 4 (2016): 725–44.

11 Interview with Tracy Hill, March 2017.

12 Interview with Gary Hera-Singh, September 2016.

13 Interview with Tracy Hill, March 2017.

14 Interview with Tracy Hill, March 2017.

15 Interview with Garry Hera-Singh, September 2016.

16 Interview with Garry Hera-Singh, September 2016.

17 Interview with Garry Hera-Singh, September 2016.

18 Interview with Garry Hera-Singh, September 2016.

19 Interview with Garry Hera-Singh, September 2016; J. H. Harris, R. T. Kingsford, W. Peirson, and L. J. Baumgartner, "Mitigating the Effects of Barriers to Freshwater Fish Migrations: The Australian Experience," *Marine and Freshwater Research* 68, no. 4 (2017): 614–28; Settre and Wheeler, "Century."

20 Settre and Wheeler, "Century."

21 Interview with Garry Hera-Singh, September 2016.

22 Interview with Garry Hera-Singh, September 2016.

23 Interview with Tracy Hill, March 2017.

24 Tracey Rogers, pers. comm., March 2019; Working Group: Long-Nosed Fur Seals in the Coorong and Lower Lakes, "Communiqué no. 15b," Department of Environment, Water and Natural Resources, Government of South Australia, February 2018.

25 Tracey Rogers, pers. comm., March 2019.

26 Shaughnessy, Goldsworthy, and Mackay, "Long-Nosed"; Chilvers and Goldsworthy, *Arctocephalus*.

27 Chilvers and Goldsworthy, *Arctocephalus*.

28 Shaughnessy, Goldsworthy, and Mackay, "Long-Nosed," 101. See also, for example, Roger Kirkwood, Robert M. Warneke, and John P. Y. Arnould, "Recolonization of Bass Strait, Australia, by the New Zealand Fur Seal, Arctocephalus forsteri," *Marine Mammal Science* 25, no. 2 (2009): 441–49; R. Kirkwood and S. Goldsworthy, *Fur Seals and Sea Lions* (Collingwood, Vic.:

CSIRO, 2013); John Ling, "Exploitation and Sea Lions from Australian, New Zealand and Adjacent Subantarctic Islands during the Eighteenth, Nineteenth and Twentieth Centuries," *Australian Zoologist* 31, no. 2 (1999): 323–50.
29 Shaughnessy, Goldsworthy, and Mackay, "Long-Nosed," 101; Ling, "Exploitation"; Michael Simmons, "Coorong Seal Issues Highlighted by MP," *The Times* (Victor Harbour), 18 June 2015, https://www.victorharbortimes.com.au/story/3155297/coorong-fishermens-incomes-on-the-line-mp/.
30 Interview with Meryl and Peter Mansfield-Cameron, March 2017.
31 Alice Dempster, "Ngarrindjeri Elder Darrell Sumner Kills Coorong Seals," *The Times*, 17 September 2015, https://www.victorharbortimes.com.au/story/3355810/ngarrindjeri-elder-darrell-sumner-kills-coorong-seals/.
32 Though the state government investigated his actions, a court case did not eventuate. Dempster, "Ngarrindjeri."
33 Interview with Meryl and Peter Mansfield-Cameron, March 2017.
34 Bell, ed., *Listen*, 26–27.
35 Sturt, *Two Expeditions*, 268.
36 Brittany Evins and Leonie Thorne, "Managing Coorong Seals Could Provide Local Jobs, Say Indigenous Elders," *ABC News*, 27 October 2019, https://www.abc.net.au/news/rural/2019-10-27/coorong-fur-seals-need-better-management/11640288.
37 Anna Tsing, *Friction: An Ethnography of Global Connection* (Princeton, NJ: Princeton University Press, 2011), 27 and 28.
38 Rosemary-Claire Collard, "Disaster Capitalism and the Quick, Quick, Slow Unravelling of Animal Life," *Antipode* 50, no. 4 (2018): 911.
39 Shaughnessy, Goldsworthy, and Mackay, "Long-Nosed," 101; Ling, "Exploitation."
40 Ling, "Exploitation"; John K. Ling, "Impact of Colonial Sealing on Seal Stocks around Australia, New Zealand and Subantarctic Islands between 150 and 170 Degrees East," *Australian Mammalogy* 24, no. 1 (2002): 117–26; Kirkwood, Warneke, and Arnould, "Recolonization"; Ngarrindjeri Nation, "Ngarrindjeri Nation Yarluwar-ruwe Plan: Caring for Ngarrindjeri Country and Culture," introduction and conclusion by Steve Hemming, in *Natural History of the Coorong, Lower Lakes, and Murray Mouth Region (Yarluwar-Ruwe)*, ed. Luke Mosely, Qifeng Ye, Scoresby Shepard, Steve Hemming, and Robert Fitzpatrick (Adelaide: University of Adelaide Press, 2018), 14.
41 Ling, "Exploitation."
42 Lynette Russell, *Roving Mariners: Australian Aboriginal Whalers and Sealers in the Southern Oceans, 1790–1870* (New York: SUNY Press, 2012); Lynette Russell, "'Dirty Domestics and Worse Cooks': Aboriginal Women's Agency and Domestic Frontiers, Southern Australia, 1800–1850," *Frontiers: A Journal of Women Studies* 28, nos. 1–2 (2007): 18–46; Lyndall Ryan, *The Aboriginal Tasmanians* (Sydney: Allen and Unwin, 1996); Lyndall Ryan, *Tasmanian Aborigines: A History since 1803* (Sydney: Allen and Unwin, 2012); Rebe Taylor, *Unearthed: The Aboriginal Tasmanians of Kangaroo Island* (Melbourne:

Wakefield Press, 2002); Kay Merry, Stephen Murray-Smith, and Iain Stuart, "The Cross-Cultural Relationships between the Sealers and the Tasmanian Aboriginal Women at Bass Strait and Kangaroo Island in the Early Nineteenth Century," *Journal of Australian Studies* 66 (2000): 73–84.

43 Ling, "Exploitation," 323.
44 Ling, "Exploitation," 327.
45 Ling, "Exploitation."
46 P. D. Shaughnessy, G. L. Shaughnessy, and L. Fletcher, "Recovery of the Fur Seal Population at Macquarie Island," *Papers and Proceedings of the Royal Society of Tasmania* 122, no. 1 (1988): 177–87.
47 Ling, "Exploitation," 339–40.
48 Ling, "Exploitation," 330–31.
49 Ling, "Exploitation," 339.
50 *South Australian Animals and Birds Protection Act* 1919; *Chronicle*, 26 April 1913, 14; *The Register*, 23 August 1918, 5; *Mercury*, 21 October 1929, 5; *Canberra Times*, 30 September 1987, 17; *Canberra Times*, 30 September 1987, 17.
51 *South Australian Fauna Conservation Act 1964*; *South Australian National Parks and Wildlife Act 1972*.
52 P. D. Shaughnessy and N. Gales, "First Survey of Fur Seals and Sea Lions in Western Australia and South Australia," *Australian Ranger Bulletin* 5 (1990): 46–49; L. J. Boren, C. G. Muller, and N. J. Gemmell, "Colony Growth and Pup Condition of the New Zealand Fur Seal (Arctocephalus forsteri) on the Kaikoura Coastline compared with Other East Coast Colonies," *Wildlife Research* 33 (2006): 497–505; Shaughnessy, Goldsworthy, and Mackay, "Long-Nosed"; Chilvers and Goldsworthy, *Arctocephalus*.
53 P. D. Shaughnessy, N. J. Gales, T. E. Dennis, and S. D. Goldsworthy, "Distribution and Abundance of New Zealand Fur Seals, *Arctocephalus forsteri*, in South Australia and Western Australia," *Wildlife Research* 21 (1994): 667–95.
54 Shaughnessy, Goldsworthy, and Mackay, "Long-Nosed," 107. Australian sea lion numbers have steadily declined, and they are currently listed as endangered. S. D. Goldsworthy, *Neophoca cinerea: The IUCN Red List of Threatened Species* (2015), e.T14549A45228341, http://dx.doi.org/10.2305/IUCN.UK.2015-2.RLTS.T14549A45228341.en.
55 Working Group, "Communiqué no. 15b."
56 Tracey Rogers, pers. comm., March 2019; Kirkwood and Goldsworthy, *Fur Seals and Sea Lions*.
57 Tracey Rogers, pers. comm., March 2019.
58 Russell, *Roving*, 100–102; Lyndall Ryan, "Aboriginal Women and Agency in the Process of Conquest: A Review of Some Recent Work," *Australian Feminist Studies* 1, no. 2 (1986): 35–43. Tracey Rogers, pers. comm., March 2019.
59 For example, *The Argus*, 15 December 1933, 5; *The Australasian*, 27 January 1934, 42; *Sunday Times*, 9 September 1919, 21.
60 Interview with Tracy Hill, March 2019.

61 Seals' oceanic predators include orcas/killer whales and some species of sharks. Kirkwood and Goldsworthy, *Fur Seals and Sea Lions*.
62 Graham J. Edgar, Trevor J. Ward, and Rick D. Stuart-S mith, "Rapid Declines across Australian Fishery Stocks Indicate Global Sustainability Targets Will Not Be Achieved without an Expanded Network of 'No-Fishing' Reserves," *Aquatic Conservation: Marine and Freshwater Ecosystems* 28, no. 6 (2018): 1337–50; Alistair J. Hobday, Gretta T. Pecl, Beth Fulton, Heidi Pethybridge, Cathy Bulman, and Cecilia Villanueva, "Climate Change Impacts, Vulnerabilities and Adaptations: Australian Marine Fisheries," in *Impacts of Climate Change on Fisheries and Aquaculture: Synthesis of Current Knowledge, Adaptation, and Mitigation Options*, ed. Manuel Barange, Tarub Bahri, Malcolm C. M. Beveridge, Kevern L. Cochrane, Siomon Funge-Smith, and Florence Poulain (Rome: Food and Agriculture Organisation, 2018), 347–62.
63 Alagona, Sandlos, and Wiersma, "Past," 50.
64 Tracey Rogers, pers. comm., March 2019.
65 Ling, "Exploitation."
66 Judith E. King, "On the Identity of the Fur Seals of Australia," *Nature* 219, no. 5154 (1968): 632–33.
67 Peter D. Shaughnessy and Simon D. Goldsworthy, "Long-Nosed Fur Seal: A New Vernacular Name for the Fur Seal, Arctocephalus forsteri, in Australia," *Marine Mammal Science* 31, no. 2 (2015): 830 and 831.
68 Shaughnessy and Goldsworthy, "Long-Nosed," 830.
69 Shaughnessy and Goldsworthy, "Long-Nosed," 830.
70 Morgan Churchill, Robert W. Boessenecker, and Mark T. Clementz, "Colonization of the Southern Hemisphere by Fur Seals and Sea Lions (Carnivora: Otariidae) Revealed by Combined Evidence Phylogenetic and Bayesian Biogeographical Analysis," *Zoological Journal of the Linnean Society* 172, no. 1 (2014): 200–225.
71 Shaughnessy and Goldsworthy, "Long-Nosed," 830.
72 Interview with Tracy Hill, March 2017.
73 Lesley Head, "Decentring 1788: Beyond Biotic Nativeness," *Geographical Research* 50, no. 2 (2012): 166–78.
74 Jane Mulcock and David Trigger, "Ecology and Identity: A Comparative Perspective of the Negotiation of 'Nativeness,'" in *Toxic Belonging: Identity and Ecology in Southern Africa*, ed. Dan Wylie (Newcastle: Cambridge Scholars Publishing, 2008): 178; David Trigger, Jane Mulcock, Andrea Gaynor, and Yann Toussaint, "Ecological Restoration, Cultural Preferences and the Negotiation of 'Nativeness' in Australia," *Geoforum* 39, no. 3 (2008): 1273–83, 1274; Chew and Hamilton, "Rise," 36; van Dooren, "Invasive."
75 This resonates with an earlier way of thinking about animals as on the same moral spectrum as people, hence, for example, all snakes are inherently evil and must be killed. Keith Thomas, *Man and the Natural World: Changing Attitudes in England 1500–1800* (London: Penguin, 1991).

76 Mark McGrouther, "European Carp, *Cyprinus carpio* Linnaeus, 1758," Australian Museum, https://australian.museum/learn/animals/fishes/european-carp-cyprinus-carpio/?gclid=CjoKCQjwpZT5BRCdARIsAGEXozkNTD2QTMiwB86AS-G-xXaneYcj1_54JD2xq9mdLASRIZlR-kYUbFoaAr1vEALw_wcB; J. D. Koehn, "Carp (*Cyprinus carpio*) as a Powerful Invader in Australian Waterways," *Freshwater Biology* 49 (2004): 882–94.

77 Interview with Tracy Hill, March 2017. See also Jennifer Atchison, Leah Gibbs, and Eli Taylor, "Killing Carp (Cyprinus carpio) as a Volunteer Practice: Implications for Community Involvement in Invasive Species Management and Policy," *Australian Geographer* 48, no. 3 (2017): 333–48.

78 Interview with Tracy Hill, March 2017.

79 Interview with Tracy Hill, March 2017.

80 Tsing, *Mushroom*.

81 Interview with Garry Hera-Singh, September 2016; Working Group, "Communiqué no. 15b."

82 Interview with Garry Hera-Singh, September 2016.

83 Anna Lowenhaupt Tsing, "Blasted Landscapes (and the Gentle Arts of Mushroom Picking)," in *The Multispecies Salon: Gleanings from a Para-site*, ed. Eben Kirksey (Durham, NC: Duke University Press, 2014), 87–109.

AFTERWORD

1 For example, Cameron Gooley, "Claims Cod Killed in Menindee Algal Bloom Were 100 Years Old Disputed by Academic," *ABC News*, 11 January 2019, https://www.abc.net.au/news/2019-01-11/fish-deaths-from-algae-bloom-tragedy-but-are-they-100/10706192; Australian Academy of Science, "Investigation of the Causes of the Mass Fish Kills in Menindee in the Summer of 2018–2019," Australian Academy of Science, 18 August 2019.

2 Jessie Davies, "Ramsar-Protected Macquarie Marshes Wetland on Fire with 90pc of Crucial Reed Bed Razed," *ABC News*, 28 October 2019, https://www.abc.net.au/news/2019-10-28/macquarie-marshes-on-fire-90pc-reed-bed-razed/11645914.

3 CSIRO and Bureau of Meteorology, "Climate Change in Australia: Technical Report 2007" (Melbourne: CSIRO, 2007).

4 CSIRO, *Water Availability in the Murray-Darling Basin*, report to the Australian Government, Canberra, 2008.

5 Interim Inspector-General of Murray-Darling Basin Water Resources, "Impact of Lower Inflows on State Shares under the Murray-Darling Basin Agreement," Australian Government, Canberra, 2020.

6 Kerrylee Rogers and Timothy J. Ralph, "Impacts of Hydrological Changes on Floodplain Wetland Biota," in *Floodplain Wetland Biota in the Murray-Darling Basin: Water and Habitat Requirements*, ed. Kerrylee Rogers and Timothy J. Ralph (Melbourne: CSIRO, 2010), 311–25.

7 Susan M. Haig, Sean P. Murphy, John H. Matthews, Ivan Arismendi, and Mohammad Safeeq, "Climate-Altered Wetlands Challenge Waterbird Use and Migratory Connectivity in Arid Landscapes," *Scientific Reports* 9, no. 1 (2019): 1–10.

8 *ABC Q&A*, 29 October 2019, transcribed in Bruce Shillingsworth, "'Why Are We Selling Water to Make Profit?,'" *Green Left Weekly*, 30 October 2019, https://www.greenleft.org.au/content/bruce-shillingsworth-why-are-we-selling-water-make-profit.

9 C. W. Rigby, Alan Rosen, H. L. Berry, and C. R. Hart, "If the Land's Sick, We're Sick," *Australian Journal of Rural Health* 19 (2011): 249–54.

10 W. Nikolakis, Q. Grafton, and A. Nygaard, "Indigenous Communities and Climate Change: A Recognition, Empowerment and Devolution (RED) Framework in the Murray-Darling Basin, Australia," *Journal of Water and Climate Change* 7, no. 1 (2016): 169–83.

SELECTED BIBLIOGRAPHY

Adams, M. J., V. Cavanagh, and B. Edmunds. "Bush Lemons and Beach Hauling: Evolving Traditions and New Thinking for Protected Areas Management and Aboriginal People in New South Wales." In *Protecting Country: Indigenous Governance and Management of Protected Areas*, edited by D. Smyth and G. R. Ward, 31–45. Canberra: AIATSIS, 2008.
Alagona, Peter S., John Sandlos, and Yolanda F. Wiersma. "Past Imperfect: Using Historical Ecology and Baseline Data for Conservation and Restoration Projects in North America." *Environmental Philosophy* 9, no. 1 (2012): 49–70.
Anderson, Warwick. *Colonial Pathologies: American Tropical Medicine, Race, and Hygiene in the Philippines*. Durham, NC: Duke University Press, 2006.
———. "Excremental Colonialism: Public Health and the Poetics of Pollution." *Critical Inquiry* 21, no. 3 (1995): 640–69.
———. "Nowhere to Run, Rabbit: The Cold-War Calculus of Disease Ecology." *History and Philosophy of the Life Sciences* 39, no. 13 (2017). https://doi.org/10.1007/s40656-017-0140-7.
Asdal, Kristin. "The Problematic Nature of Nature: The Post-constructivist Challenge to Environmental History." *History and Theory* 42, no. 4 (2003): 60–74.
Barua, M. "Lively Commodities and Encounter Value." *Environment and Planning D: Society and Space* 34, no. 4 (2016): 725–44.
Bawaka Country, including L. Burarrwanga, R. Ganambarr, M. Ganambarr-Stubbs, B. Ganambarr, D. Maymuru, S. Wright, S. Suchet-Pearson, K. Lloyd and J. Sweeney. "Co-becoming Time/s: Time/s-as-telling-as-time/s." In *Methodological Challenges in Nature-Culture and Environmental History Research*, edited by J. Thorpe, S. Rutherford, and L. A. Sandberg, 101–12. London: Routledge, 2017.
Beattie, James. *Empire and Environmental Anxiety: Health, Science, Art and Conservation in South Asia and Australasia, 1800–1920*. London: Springer, 2011.
Bell, Diane. *Ngarrindjeri Wurruwarrin: A World That Is, Was, and Will Be*. North Geelong, Vic.: Spinifex Press, 1998.

———, ed. *Listen to Ngarrindjeri Women Speaking*. Melbourne: Spinifex Press, 2008.

Benson, Etienne. "Animal Writes: Historiography, Disciplinarity, and the Animal Trace." In *Making Animal Meaning*, edited by L. Kalof and G. M. Montgomery, 3–16. East Lansing: Michigan State University Press, 2011.

Biehler, Dawn Day, and Gregory L. Simon. "The Great Indoors: Research Frontiers on Indoor Environments as Active Political-Ecological Spaces." *Progress in Human Geography* 35, no. 2 (2011): 172–92.

Biggs, David. *Quagmire: Nation-Building and Nature in the Mekong Delta*. Seattle: University of Washington Press, 2012.

Boardman, Robert. *International Organization and the Conservation of Nature*. London: Macmillan, 1981.

———. *The International Politics of Bird Conservation: Biodiversity, Regionalism and Global Governance*. Cheltenham: Edward Elgar, 2006.

Boileau, Joanna. *Chinese Market Gardening in Australia and New Zealand: Gardens of Prosperity*. London: Palgrave Macmillan, 2017.

Bowker, Geoffrey C., and Susan Leigh Star. *Sorting Things Out: Classification and Its Consequences*. Cambridge, MA: MIT Press, 2000.

Braun, Bruce. *The Intemperate Rainforest: Nature, Culture, and Power on Canada's West Coast*. Minneapolis: University of Minnesota Press, 2002.

Brockington, Dan. *Fortress Conservation: The Preservation of the Mkomazi Game Reserve, Tanzania*. Bloomington: Indiana University Press, 2002.

Butterly, Lauren. "'For the Reasons Given in Akiba . . .': Karpany V Dietman [2013] Hca 47." *Indigenous Law Bulletin* 8, no. 10 (2014): 23–26.

Candiani, Vera S. *Dreaming of Dry Land: Environmental Transformation in Colonial Mexico City*. Stanford, CA: Stanford University Press, 2014.

Chew, Matthew K., and Andrew L. Hamilton. "The Rise and Fall of Biotic Nativeness: A Historical Perspective." In *Fifty Years of Invasion Biology: The Legacy of Charles Elton*, edited by David Richardson, 35–47. Oxford: Blackwell, 2011.

Cioc, Mark. *The Game of Conservation: International Treaties to Protect the World's Migratory Animals*. Athens: Ohio University Press, 2009.

Clarke, Philip A. "Birds as Totemic Beings and Creators in the Lower Murray, South Australia." *Journal of Ethnobiology* 36, no. 2 (2016): 277–94.

Coates, Peter. *American Perceptions of Immigrant and Invasive Species: Strangers on the Land*. Berkeley: University of California Press, 2007.

Collard, Rosemary-Claire. "Disaster Capitalism and the Quick, Quick, Slow Unravelling of Animal Life." *Antipode* 50, no. 4 (2018): 910–28.

Davidson, N. C. "How Much Wetland Has the World Lost? Long-term and Recent Trends in Global Wetland Area." *Marine and Freshwater Research* 65 (2014): 934–41.

Davies, P., and S. Lawrence. *Sludge: Disaster on Victoria's Goldfields*. Melbourne: Black, 2019.

Douglas, Mary. *Purity and Danger: An Analysis of Concepts of Pollution and Taboo*. 1966; repr. London: Routledge, 2003.

Dow, Coral. "'Sportsman's Paradise': The Effects of Hunting on the Avifauna of the Gippsland Lakes." *Environment and History* 14 (2008): 145–64.

Dunlap, Thomas R. *Nature and the English Diaspora: Environment and History in the United States, Canada, Australia, and New Zealand.* Cambridge: Cambridge University Press, 1999.

Eriksen, C., and D. L. Hankins. "Colonisation and Fire: Gendered Dimensions of Indigenous Fire Knowledge Retention and Revival." In *The Routledge Handbook of Gender and Development*, edited by A. Coles, L. Gray and J. Momsen, 129–37. New York: Routledge, 2015.

Erwin, Kevin L. "Wetlands and Global Climate Change: The Role of Wetland Restoration in a Changing World." *Wetlands Ecology and Management* 17, no.1 (2009): 71–84.

Finlayson, C. M., and D. S. Mitchell. "Australian Wetlands: The Monitoring Challenge." *Wetlands Ecology and Management* 7, nos. 1–2 (1999): 105–12.

Finlayson, C. M., Jennifer Ann Davis, Peter A. Gell, R. T. Kingsford, and K. A. Parton. "The Status of Wetlands and the Predicted Effects of Global Climate Change: The Situation in Australia." *Aquatic Sciences* 75, no. 1 (2013): 73–93.

Gammage, William B. *The Biggest Estate on Earth: How Aborigines Made Australia.* Sydney: Allen and Unwin, 2011.

———. "The Wiradjuri War, 1838–40." *The Push* 16 (1983): 3–17.

Garone, Phillip. *The Fall and Rise of the Wetlands of California's Great Central Valley.* Berkeley: University of California Press, 2011.

Goodall, Heather. "Exclusion and Re-emplacement: Tensions around Protected Areas in Australia and Southeast Asia." *Conservation and Society* 4, no. 3 (2006): 383–95.

———. "Landscapes of Meaning: The View from Within the Indian Archipelago." *Transforming Cultures eJournal* 3, no. 1 (2008).

Goodall, Heather, and Allison Cadzow. *Rivers and Resilience: Aboriginal People on Sydney's Georges River.* Sydney: University of New South Wales Press, 2009.

Graham, Mary. "Some Thoughts about the Philosophical Underpinnings of Aboriginal Worldviews." *Worldviews: Environment, Culture, Religion* 3 (1999): 105–18.

Greer, Kirsten. "Geopolitics and the Avian Imperial Archive: The Zoogeography of Region-Making in the Nineteenth-Century British Mediterranean." *Annals of the Association of American Geographers* 103, no. 6 (2013): 1317–31.

Griffiths, Tom. *Forests of Ash: An Environmental History.* Cambridge: Cambridge University Press, 2001.

Grosz, Elizabeth A. *Volatile Bodies: Toward a Corporeal Feminism.* Bloomington: Indiana University Press, 1994.

Halliday, Stephen. "Death and Miasma in Victorian London: An Obstinate Belief." *BMJ: British Medical Journal* 323, no. 7327 (2001): 1469–71.

Hamby, Louise, and Doreen Mellor. "Fibre Tracks." In *Oxford Companion to Aboriginal Art and Culture*, edited by Margo Neale, Sylvia Kleinert, and Robyne Bancroft, 371–72. Oxford: Oxford University Press, 2000.

Hamilton, Sarah R. *Cultivating Nature: The Conservation of a Valencian Working Landscape*. Seattle: University of Washington Press, 2018.

Haraway, Donna. "Situated Knowledges: The Science Question in Feminism and the Privilege of Partial Perspective." *Feminist Studies* 14, no. 3 (1988): 575–99.

———. *When Species Meet*. Minneapolis: University of Minnesota Press, 2008.

Hartwig, Lana, Sue Jackson, and Natalie Osborne. "Recognition of Barkandji Water Rights in Australian Settler-Colonial Water Regimes." *Resources* 7, no. 1 (2018): 16.

Hattam, R., D. Rigney, and S. Hemming. "Reconciliation? Culture, Nature and the Murray River." In *Fresh Water: New Perspectives on Water in Australia*, edited by E. Potter, A. Mackinnon, S. McKenzie, and J. McKay, 105–22. Carlton, Vic.: Melbourne University Press, 2007.

Head, Lesley. "Decentring 1788: Beyond Biotic Nativeness." *Geographical Research* 50, no. 2 (2012): 166–78.

———. *Second Nature: The History and Implications of Australia as Aboriginal Landscape*. Syracuse: Syracuse University Press, 2000.

Hemming, S., and D. Rigney. "Ngarrindjeri Futures: Negotiating a Future through Caring for Ruwe/Ruwar (Lands, Waters and All Living Things)." In *Innovation for 21st Century Conservation*, edited by P. Figgis, J. Fitzsimons and J. Irving, 186–91. Sydney: Australian Committee for IUCN, 2012.

———. "Restoring Murray Futures: Incorporating Indigenous Knowledge, Values and Interests into Environmental Water Planning in the Coorong and Lakes Alexandrina and Albert Ramsar Wetland." Goyder Institute for Water Research Technical Report Series No. 16/8, Adelaide, South Australia, 2016.

Hill, Sarah H. *Weaving New Worlds: Southeastern Cherokee Women and Their Basketry*. Chapel Hill: University of North Carolina Press, 1997.

Humphries, Paul. "Historical Indigenous Use of Aquatic Resources in Australia's Murray-Darling Basin, and Its Implications for River Management." *Ecological Management and Restoration* 8, no. 2 (2007): 106–13.

Hutton, Drew, and Libby Connors. *History of the Australian Environment Movement*. Cambridge: Cambridge University Press, 1999.

Ingold, Tim. *The Perception of the Environment: Essays on Livelihood, Dwelling and Skill*. London: Routledge, 2002.

Jackson, Sue, and Marcia Langton. "Trends in the Recognition of Indigenous Water Needs in Australian Water Reform: The Limitations of 'Cultural' Entitlements in Achieving Water Equity." *Journal of Water Law* 22, nos. 2–3 (2011): 109–23.

Jackson, Sue, and Bradley Moggridge. "Indigenous Water Management." *Australasian Journal of Environmental Management* 6, no. 3 (2019): 193–96.

Jackson, Sue, Carmel Pollino, Kirsten Maclean, Rosalind Bark, and Bradley Moggridge. "Meeting Indigenous Peoples' Objectives in Environmental Flow Assessments: Case Studies from an Australian Multi-jurisdictional Water Sharing Initiative." *Journal of Hydrology* 522 (2015): 141–51.

Jacobs, Nancy. *Birders of Africa: History of a Network*. New Haven, CT: Yale University Press, 2016.

Jarman, Peter, and Margaret Brock. "The Evolving Intent and Coverage of Legislation to Protect Biodiversity in New South Wales." In *Threatened Species Legislation: Is It Just an Act?*, edited by Pat Hutchings, Daniel Lunney, and Chris Dickman, 1–19. Mosman: Royal Zoological Society of New South Wales, 2004.

Jerome, Paddy. "Boobarran Ngummin: The Bunya Mountains." *Queensland Review* 9, no. 2 (2002): 1–5.

Ju, Xiaotang, Fusuo Zhang, Xuemei Bao, V. Römheld, and M. Roelcke. "Utilization and Management of Organic Wastes in Chinese Agriculture: Past, Present and Perspectives." *Science in China Series C: Life Sciences* 48, no. 2 (2005): 965–79.

Kabaila, Peter Rimas. *Wiradjuri Places: The Murrumbidgee River Basin with a Section on Ngunawal Country*. Jamison Centre: Black Mountain Projects, 1995.

Kiechle, Melanie A. *Smell Detectives: An Olfactory History of Nineteenth-Century Urban America*. Seattle: University of Washington Press, 2017.

Kimmerer, R. W. *Braiding Sweetgrass: Indigenous Wisdom, Scientific Knowledge and the Teachings of Plants*. Minneapolis: Milkweed Editions, 2013.

Kingsford, Richard T., Alberto Basset, and Leland Jackson. "Wetlands: Conservation's Poor Cousins." *Aquatic Conservation: Marine and Freshwater Ecosystems* 26, no. 5 (2016): 892–916.

Langford-Smith, Trevor, and Ian Rutherford. *Water and Land: Two Case Studies in Irrigation*. Canberra: Australian National University Press, 1966.

Langston, Nancy. *Where Land and Water Meet: A Western Landscape Transformed*. Seattle: University of Washington Press, 2005.

Langton, Marcia. "Arts, Wilderness and Terra Nullius." In *Ecopolitics IX: Perspectives on Indigenous People's Management of Environment Resources*, edited by R. Sultan. Darwin: Northern Land Council, 1995.

———. *Well, I Heard It on the Radio and I Saw It on the Television*. Vol. 9. Sydney: Australian Film Commission, 1993.

Latour, Bruno. *We Have Never Been Modern*. Cambridge, MA: Harvard University Press, 1993.

Lewis, Michael. *Inventing Global Ecology: Tracking the Biodiversity Ideal in India, 1947–1997*. Athens: Ohio University Press, 2004.

Ling, John, "Exploitation and Sea Lions from Australian, New Zealand and Adjacent Subantarctic Islands during the Eighteenth, Nineteenth and Twentieth Centuries." *Australian Zoologist* 31, no. 2 (1999): 323–50.

Lloyd, C. J. *Either Drought or Plenty: Water Development and Management in NSW*. Sydney: NSW Department of Water Resources, 1988.

McClure, H. Elliott. "Migratory Animal Pathological Survey: Progress Report 1967." Washington, DC: US Army, 1968.

McNeill, J. R. *Mosquito Empires: Ecology and War in the Greater Caribbean, 1620–1914*. Cambridge: Cambridge University Press, 2010.

Mitchell, Timothy. *Rule of Experts: Egypt, Techno-politics, Modernity*. Berkeley: University of California Press, 2002.

Mitman, Gregg. "Living in a Material World." *Journal of American History* 100 (June 2013): 128–30.

Moggridge, Bradley. "Where Is the Aboriginal Water Voice through the Current Murray-Darling Crisis?" *Irrigation Australia: The Official Journal of Irrigation Australia* 34 (2018): 34–35.

Moggridge, Bradley J., Lyndal Betterridge, and Ross M. Thompson. "Integrating Aboriginal Cultural Values into Water Planning: A Case Study from New South Wales, Australia." *Australasian Journal of Environmental Management* 26, no. 3 (2019): 273–86.

Moorcroft, Heather. "Paradigms, Paradoxes and a Propitious Niche: Conservation and Indigenous Social Justice Policy in Australia." *Local Environment* 21, no. 5 (2016): 591–614.

Morgan, Ruth. *Running Out? Water in Western Australia*. Perth: University of Western Australia Press, 2015.

Muir, Cameron. *The Broken Promise of Agricultural Progress: An Environmental History*. London: Routledge, 2014.

Mulcock, Jane, and David Trigger. "Ecology and Identity: A Comparative Perspective of the Negotiation of 'Nativeness.'" In *Toxic Belonging: Identity and Ecology in Southern Africa*, edited by Dan Wylie, 178–98. Newcastle, UK: Cambridge Scholars, 2008.

Mulligan, Martin, and Stuart Hill. *Ecological Pioneers: A Social History of Australian Thought and Action*. Cambridge: Cambridge University Press, 2001.

Nance, Susan, ed. *The Historical Animal*. Syracuse, NY: Syracuse University Press, 2015.

Nash, Linda. *Inescapable Ecologies: A History of Environment, Disease, and Knowledge*. Berkeley: University of California Press, 2006.

Ngarrindjeri Nation. "Ngarrindjeri Nation Yarluwar-ruwe Plan: Caring for Ngarrindjeri Country and Culture." Introduction and conclusion by Steve Hemming. In *Natural History of the Coorong, Lower Lakes, and Murray Mouth Region (Yarluwar-Ruwe)*, edited by Luke Mosely, Qifeng Ye, Scoresby Shepard, Steve Hemming, and Robert Fitzpatrick, 3–20. Adelaide: University of Adelaide Press, 2018.

Pascoe, Bruce. *Dark Emu Black Seeds: Agriculture or Accident?* Broome, Western Australia: Magabala Books, 2014.

Pittock, James. and C. M. Finlayson. "Climate Change Adaptation in the Murray-Darling Basin: Reducing Resilience of Wetlands with Engineering." *Australasian Journal of Water Resources* 17, no. 2 (2013): 161–69.

Poirier, Robert, and David Ostergren. "Evicting People from Nature: Indigenous Land Rights and National Parks in Australia, Russia, and the United States." *Natural Resources Journal* 42, no. 2 (Spring 2002): 331–51.

Powell, Joseph Michael. *Watering the Garden State: Water, Land and Community in Victoria, 1834–1988*. Sydney: Allen and Unwin, 1989.

Pratt, Mary Louise. "Arts of the Contact Zone." *Profession* (1991): 33–40.

Ralph, Timothy J., Paul Hesse, and Tsuyoshi Kobayashi. "Wandering Wetlands: Spatial Patterns of Historical Channel and Floodplain Change in the

Ramsar-Listed Macquarie Marshes, Australia." *Marine and Freshwater Research* 67, no. 6 (2016): 782–802.

Rigby, C. W., Alan Rosen, H. L. Berry, and C. R. Hart. "If the Land's Sick, We're Sick." *Australian Journal of Rural Health* 19 (2011): 249–54.

Ritvo, Harriet. *The Animal Estate: The English and Other Creatures in the Victorian Age*. Cambridge, MA: Harvard University Press, 1987.

Robin, Libby. "Birds and Environmental Management in Australia 1901–2001." *Australian Journal of Environmental Management* 8, no. 2 (2001): 105–13.

———. *How a Continent Created a Nation*. Sydney: University of New South Wales Press, 2007.

Rogers, Kerrylee, and Timothy J. Ralph. "Impacts of Hydrological Changes on Floodplain Wetland Biota." In *Floodplain Wetland Biota in the Murray-Darling Basin: Water and Habitat Requirements*, edited by Kerrylee Rogers and Timothy J. Ralph, 311–25. Melbourne: CSIRO, 2010.

Rose, Deborah Bird. "Land Rights and Deep Colonising: The Erasure of Women." *Aboriginal Law Bulletin* 3, no. 85 (1996): 6–13.

———. *Reports from a Wild Country: Ethics for Decolonization*. Sydney: University of New South Wales Press, 2004.

———. "What If the Angel of History Were a Dog?" *Cultural Studies Review* 12, no. 1 (2006): 67–78.

Ross, Helen, Chrissy Grant, Cathy J. Robinson, Arturo Izurieta, Dermot Smyth, and Phil Rist. "Co-management and Indigenous Protected Areas in Australia: Achievements and Ways Forward." *Australasian Journal of Environmental Management* 16, no. 4 (2009): 242–52.

Russell, Lynette. *Roving Mariners: Australian Aboriginal Whalers and Sealers in the Southern Oceans, 1790–1870*. New York: SUNY Press, 2012.

Ryan, John Charles, and Li Chen, eds. *Australian Wetland Cultures: Swamps and the Environmental Crisis*. London: Lexington Books, 2019.

Ryan, Lyndall. *Tasmanian Aborigines: A History since 1803*. Sydney: Allen and Unwin, 2012.

Shapin, Steven, and Simon Schaffer. *Leviathan and the Air-Pump: Hobbes, Boyle, and the Experimental Life*. Princeton, NJ: Princeton University Press, 2011.

Shaughnessy, Peter D., and Simon D. Goldsworthy. "Long-Nosed Fur Seal: A New Vernacular Name for the Fur Seal, Arctocephalus Forsteri, in Australia." *Marine Mammal Science* 31, no. 2 (2015): 830–32.

Sinclair, Paul. *The Murray: A River and Its People*. Melbourne: Melbourne University Press, 2001.

Sörlin, Sverker, and Paul Warde. "The Problem of the Problem of Environmental History: A Re-reading of the Field." *Environmental History* 12, no. 1 (2007): 107–30.

Stubbs, B. J. "From 'Useless Brutes' to National Treasures: A Century of Evolving Attitudes towards Native Fauna in New South Wales, 1860s to 1960s." *Environment and History* 7 (2001): 23–56.

Sturt, Charles. *Two Expeditions into the Interior of Southern Australia*. London: Smith, Edler, 1833.
Sutter, Paul. "Nature's Agents or Agents of Empire? Entomological Workers and Environmental Change during the Construction of the Panama Canal." *Isis* 98, no. 4 (2007): 724–54.
———. "The World with Us: The State of American Environmental History." *Journal of American History* 100, no. 1 (2013): 94–119.
Taylor, Rebe. *Unearthed: The Aboriginal Tasmanians of Kangaroo Island*. Melbourne: Wakefield Press, 2002.
Trevorrow, Ellen, Yvonne Koolmatrie, Doreen Kartinyeri, and Diana Wood Conroy. "Binding the Rushes: Survival of Culture." In *The Oxford Companion to Aboriginal Art and Culture*, edited by Margo Neale, Sylvia Kleinert, and Robyne Bancroft, 371–72. Oxford: Oxford University Press, 2000.
Trigger, David, Jane Mulcock, Andrea Gaynor, and Yann Toussaint. "Ecological Restoration, Cultural Preferences and the Negotiation of 'Nativeness' in Australia." *Geoforum* 39, no. 3 (2008): 1273–83.
Tsing, Anna. *Friction: An Ethnograhy of Global Connection*. Princeton, NJ: Princeton University Press, 2011.
———. *The Mushroom at the End of the World: On the Possibility of Life in Capitalist Ruins*. Princeton, NJ: Princeton University Press, 2015.
Tunstall, Elizabeth Dori. "Be Rooted: Learning from Aboriginal Dyeing and Weaving." *The Conversation*, 11 August 2015. https://theconversation.com/be-rooted-learning-from-aboriginal-dyeing-and-weaving-45940.
van Dooren, Thom, Eben Kirksey, and Ursula Münster. "Multispecies Studies: Cultivating Arts of Attentiveness." *Environmental Humanities* 8, no. 1 (2016): 1–23.
Vileisis, Ann. *Discovering the Unknown Landscape: A History of America's Wetlands*. Washington, DC: Island Press, 1999.
Warde, Paul, Libby Robin, and Sverker Sörlin. *The Environment: A History of the Idea*. Baltimore: Johns Hopkins University Press, 2018.
Watson, Irene. "Buried Alive." *Law and Critique* 13, no. 3 (2002): 253–69.
Weir, Jessica. *Murray River Country: An Ecological Dialogue with Traditional Owners*. Canberra: Aboriginal Studies Press, 2009.
Weir, Jessica, D. R. J. Crew, and J. L. Crew. "Wetland Forest Culture: Indigenous Activity for Management Change in the Southern Riverina, New South Wales." *Australasian Journal of Environmental Management* 20, no. 3 (2013): 193–207.
West, Paige, James Igoe, and Dan Brockington. "Parks and Peoples: The Social Impact of Protected Areas." *Annual Review of Anthropology* 35 (2006): 251–77.
Wilson, Robert. *Seeking Refuge: Birds and Landscapes of the Pacific Flyway*. Seattle: University of Washington Press, 2010.

INDEX

Note: Page numbers in *italic* indicate a figure on the corresponding page.

Aboriginal Lands Trust, 30
Aboriginal Water Initiative, 41
absences and presences, 181–87
abuse, 181–82
access: Aboriginal people's access to water, 13, 196; and bird protection legislation, 135, 158; and cattle grazing, 24; and conflict, 50; and weaving, 20, 26, 35, 41–45
acclimatization, 131
Adelaide, 126–27, 129, 135, 137
Africa, 65, 78, 131, 147
agencies, 7, 10–11, 15, 17–18, 20, 144, 196
agency, 10, 18, 48, 55, 66, 144
agriculture, 3, 6–7, 10–11, 13–14, 169, 194, 196–97; and Chinese migrants, 59–62; civilizing influence of, 89; and conservation science, 115–19; cotton, 6, 23–25, 35, 41; and disease, 73, 77, 79, 81–87, 94, 96; Experiment Farm, 83; industrialized farming, 144; market gardens, 14, 61–62, 72; and migratory birds, 146, 166–67; ornamental fish farming, 190; and pelicans, 127, 133–34, 139–40; pig farming, 61; and swamps, 100–104; and weaving, 36, 40–42. *See also* rice farming
Ainu, 159
Alagona, Peter, 186
alfalfa, 83
Alligator River, 163

Anderson, Warwick, 62–63, 76
Angas, George French, 37
Anglo-Celtic colonists, 101, 111, 134
Animals and Birds Protection Act 1919, 138
Animals Protection Act 1912, 135, 138
Anopheles mosquito (*Anopheles annulipes*), 68–69, 74–75, 77–80, 79; and agriculture, 83–91; and the home, 91–95; and the troubling of boundaries, 80–83. *See also* mosquitoes
Anthropocene, 12, 21, 172–73, 180, 193
aquifers, 7, 14, 46–49, 58, 60, 66, 71
Arctocephalus forsteri, 187–89. *See also* long-nosed fur seal
assimilation, 33, 137
Australasian bitterns (*Botaurus poiciloptilus*), 100, 118, 164–65
Australian Army, 79
Australian Capital Territory, 6
Australian Committee on Waterbirds, 160
Australian fur seal (*Arctocephalus pusillus*), 170, 188–89
Australian Museum, 106–7
Australian National Water Initiative, 39
Australian pelican (*Pelecanus conspicillatus*). *See* pelicans
Australian sea lion (*Neophoca cinerea*), 170, 188
Australian shelduck (mountain duck, *Tadorna tadornoides*), 104–5

Australian wood duck (maned goose, *Chenonetta jubata*), 104–5, 107, 115
Ayres, F. G., 127, 129–30

bacteria, 48, 76, 194
Barmah Forest, 163
Barmah Forest virus, 77
barrages, 6, 14, 38, 139; and seals, 3–4, 169–71, 175–78, 184–85, 192
Barren Box Swamp, 101
Barrett, Charles, 136
Bartley, Nehemiah, 52
baselines, 171, 173, 178–79, 181–87, 190, 197
Basset Hull, A., 112
Bell, Diane, 29, 36
belonging, 77, 100, 105–6, 109, 172–73, 184, 187–93
billabongs, 148, 161
biodiversity, 11–12, 102, 117–18, 147, 166; Commonwealth Environmental Protection and Biodiversity Conservation Act 1999, 178
bird-centrism, 142–43, 167
BirdLife Australia, 118
Bird Protection Act 1900, 128, 134–35, 137
Bird Rose, Deborah, 29, 133
birds: Australasian bitterns, 100, 118, 164–65; black swans, 36, 97, 131, 136–38; cockatoos, 155; cormorants, 128–29, 134; cranes, 107, 163; crows, 107, 125, 136; galahs, 107; gannets, 127; great auks, 130; herons, 107, 127; ibis, 71–72, 97, 107; Latham's snipe, 141, 143, 145, 151–55, 157, 165; magpie, 125; magpie geese, 151–52; pheasants, 110, 131; quail, 108; silver gulls, 127; sparrows, 107; spoonbills, 107; terns, 127; water hens, 5, 97; weaver birds, 27–28. *See also* ducks; migratory birds; pelicans; waterbirds
Birds Protection Act 1881, 110
bitterns. *See* Australasian bitterns
black box–lignum vegetation, 101–2
black swans, 36, 97, 131, 136–38
Boardman, Robert, 147
bony herring, 125, 129
boundaries, 16, 19, 21; and containment of swamps, 47–48, 65, 67, 70; and disease, 77, 80–83, 94; and ducks, 98–99, 98, 105, 118. *See also* crossing
bounties, 109, 126–31, 133, 136, 139
Braun, Bruce, 142–43
Brisbane, 52–53, 55–56, 63, 68–69, 87–88

Britain, 14, 52, 54–55, 129, 131–35, 148–49; advocacy groups in, 134, 146; bird-banding program, 156; British Society for the Protection of Birds, 134; game laws, 132, 135; models of bird behavior, 154; traditions of protection, 146. *See also* colonialism; colonization
British Empire, 131, 183
British Game Laws, 132, 135
British Society for the Protection of Birds, 134
bubonic plague, 65
Budjiti, 196
Bunya Mountains, 49, 64
bureaucracies, 14–15, 29, 38, 41–43, 198
Burketown, 78
Burma, 87
Burnett, James Charles, 50
Burrendong Dam, 25–26, 40, 195
Burrinjuck Dam, 97
Burton Cleland, John, 83
bush coastal Olearia (*kengk muldi*; smoking bush), 29

California, 16, 76, 105
Cameron, Alfred, 30
Campbell, A. J., 153–54
Canada, 131, 134–35, 142
canalization, 139, 145
capitalism, 21, 169, 172–81, 184, 192, 196
carbon dioxide, 11–12
carbon sinks, 11–12, 196
carp gudgeon (*Hypseleotris* spp.), 67
Cass, "Moss," 149, 152
catadromous fish, 175–76
cattle: and Aboriginal people's access to sedges and reeds, 23–24, 24, 30, 42; and colonization, 13, 49, 52–53; shaping swamps, 55, 57, 66, 86, 99, 103. *See also* grazing
cesspits, 58–64
channelization, 13, 46, 71
Chapman, F. R. H., 127, 139
China: migrants from, 14, 59, 61–62, 64; migratory bird agreements with, 163
cholera, 58, 61
Cilento, R. W., 82
cisterns, 65–67
Clarke, Philip, 138
class, 9, 47–48; and bird protection, 131–32, 140; and disease, 59–61; and drainage, 56; and private property, 123; and the

wilderness approach of government conservation, 149
cleanliness. *See* hygiene and cleanliness
climate change, 10–12, 35–36, 140, 164, 195–97
coastal wetlands, 11–12, 163, 195
Cobourg Peninsula, 78; Wildlife Reserve and Sanctuary, 151
cockatoos, 155
cocreation, 7, 9, 13, 27–28; of changing landscapes, 99; and disease, 65; of human lives, 76; of imagined ecologies, 74, 142; of more-than-human histories, 172; of more-than-human worlds, 48, 198; of wetlands, 198
Cold War, 94, 144, 152, 156
Coleambally Irrigation Area, 5
Collard, Rosemary-Claire, 173, 180
colonialism, 9, 29, 32, 37, 42–43; colonial extractivism, 180; colonial laws, 110, 133; colonial records, 125, 179; and containment of swamps, s47–50, 52, 54–57, 62–63; ideas of wilderness and nature, 150; ideologies of racial hierarchy, 153; the language of discovery, 154; and malaria, 78; and pelicans, 136; and seals, 178, 184
colonization, 9–10, 13, 21, 197; and capitalism, 169, 180; and changes to wetlands, 102–3; and containment of swamps, 42, 54, 71; and disease, 78, 126; and environmental protection, 131, 133, 148; and hygiene, 88; and migratory birds, 153, 159–60; and pelicans, 136; and race, 137; and seals, 184–85, 189; and weaving, 27–33
comanagement, 138, 166. *See also* Kakadu National Park
Commonwealth Environmental Protection and Biodiversity Conservation Act 1999, 178
Commonwealth Scientific and Industrial Research Organisation (CSIRO), 114–15, 145, 150–52, 159–60, 162
Condon, H. T., 139
conservationism, 5, 7, 11, 15, 21, 43; and ducks, 100, 103, 111–12; and pelicans, 121, 140; and migratory birds, 142, 147–48, 150–52, 158–59, 160–61, 163–67; and seals, 178, 182–84, 187, 193. *See also* conservation science; government conservation
conservation science, 115–19, 147, 167
Convention on Biological Diversity, 164
Cooling, L. E., 83–85
Coorong and Lower Lakes Fishery, 170, 175

Coorong lagoon, 3–8, *8*, 10, 19, 21; and pelicans, 120–21, *121*, 123–28, 132, 135–39; and seals, 168–77, 179–81, 184–87, 189, 191–92; and weaving, 27–30, 35–38
Coorong National Park, 138, 167
cormorants, 128–29, 134
cotton farming, 6, 23–25, 35, 41
Council for Scientific and Industrial Research (CSIR), 87, 90, 115–16
cranes, 107, 163
crayfish, 107
creation, stories of, 25, 124–25, 132, 179
Crew, David, 167
Crew, Jeanette, 167
crimson-spotted sunfish (*Melanotaenia duboulayi*), 67
crocodiles, 147
Crompton, R., 134–35
crossing, 20, 21, 97–100; conflicting interests, 111–14; ducks and rice, 104–9, 111–15; native pests, the economic value of birds, and hunting, 109–11; postwar ecology and conservation science, 115–19; swamps in an irrigation landscape, 100–4
crown land, 53, 58, 132, 135–36, 138
Crown Lands Office, 135–36
crows, 107, 125, 136
CSIR. *See* Council for Scientific and Industrial Research
CSIRO. *See* Commonwealth Scientific and Industrial Research Organisation
Culex mosquito (*Culex fatigans*; *Culex quinquefasciatus*), 65–67, 69. *See also* mosquitoes
cultural flows, 39–40, 164
cultural knowledge, 33–35
Cumbungi reeds/rushes, 26, 49, 102
Cumpston, John H. L., 80
Curtin, Alison, 105, 117–19

Dabb, G. P. M., 157
Daily Herald, 127, 130, 137–38
Daily Telegraph, 89
dairying, 83
dams, 6, 14, 39, 73, 86, 144–46, 150; and ducks, 97, 99, 113, 116–17, 119. *See also* Burrendong Dam; Burrinjuck Dam; Hume Dam
Darling Downs, the, 49–52, 64, 69
Darling River, 5, 35, 181, 186, 194
DDT, 70, 76, 93–95, 102
decolonizing, 13, 29
dengue fever, 65, 67–70, 88, 92, 93

INDEX | 253

Department of Public Health, 65, 82–83, 85–86, 92, 95
Department of the Environment and Conservation, 149–50
Depression, the, 70
diadromous fish, 175–76
dingoes, 133, 140
diphtheria, 53, 59, 63
discovery, language of, 153–54
disease, 9, 11, 13, 46, 48; bubonic plague, 65; cholera, 58, 61; colonialist disease prevention, 62–63; dengue fever, 65, 67–70, 88, 92, 93; diphtheria, 53, 59, 63; dysentery, 53–54, 58–59; encephalitis, 77, 156; liver fluke, 112–13; measles, 63; myxomatosis, 95; scarlet fever, 53; smallpox, 126; typhoid, 53–55, 58–61, 63–64; yellow fever, 65, 69. *See also* malaria
disease ecology, 75–76
drainage, 6, 11, 101, 103, 176; and containment of swamps, 47–48, 55–59, 63, 68–70; and disease, 81, 84, 87–89; and migratory birds, 143–46, 155, 160, 163
Drayton, 50–53
Drayton Swamp. *See* Toowoomba swamps
drinking water, 58, 196
drought, 11–12, 14–16, 60, 73, 134, 172, 195; and ducks, 105–6, 108, 110, 113; Federation Drought, 80; Millennium Drought, 5, 118, 175, 190–92, 196; and weaving, 25, 35, 42
drying, 11–12, 22–23, 25–26, 38, 104, 111, 156, 194–96. *See also* droughts
ducks, 16, 131, 151, 155, 177; Australian shelduck, 104–5; Australian wood duck, 104–5, 107, 115; and ecology and conservation science, 116–19; grey teal, 104–7; and hunting, 110, 118; Pacific black duck, 104–7; pink-eared duck, 104–5, 107; and rice growing, 16, 97–100, 98, 104–9, 111–15, 117–19; whistling duck, 104–5
Dunlap, Thomas, 108, 131
Dust Bowl, 116
dysentery, 53–54, 58–59

East Asian–Australasian Flyway, 157
East Asian Flyway, 157
East Creek, 46, 71
ecology, field of, 75–76, 108, 115–19. *See also* disease ecology; landscape ecology; wetlands ecology
economics: and drought, 175; and ecology, 116; economic value of birds, 106, 109–11, 134; and fishing, 190–92; and rice farming, 112; and seals, 170. *See also* Depression, the; recession
egg collecting, 123, 132–33, 136–38, 153, 157–59
Elders, 28–29, 71, 179; Aunty Dorrie Katinyeri, 37; Darrell Sumner, 179; Paddy Jerome, 64; Peter Mansfield-Cameron, 179
Emu (journal), 136
encephalitis, 77, 156
enclosing, 4, 20, 21, 120–22
England, 54–55, 129. *See also* Britain
entomology, 9, 69, 85–91, 94–95, 101, 115–16
environmental crisis, 146–49
environmental history, 17–18, 131, 198
environmental humanities, 16–18, 198
environmentalism and environmental movement, 143–46, 148, 152, 184, 193; critical, 143
ephemeral wetlands, 161–63
Ertz imagery, 161
estuaries, 6, 151–52, 161–62, 174–76, 184–85, 187
Europe, 83, 146–49, 190
European carp (*Cyprinus carpio*), 10, 21, 172, 190–92
Ewens, M. C., 137
extinction, 130, 139, 152, 157
extractivism, 180

farming. *See* agriculture
Fauna Conservation Act, 184
federation, 80, 122, 130, 134
Federation Drought, 80
feral pigs, 42
Ferguson, Eustace, 85–86
fertilizer, 62, 191
filaria, 65, 69
Finlayson, Max, 164–65
fire, 32–33, 125, 194
fish: bony herring, 125, 129; carp gudgeon, 67; crimson-spotted sunfish, 67; European carp, 10, 21, 172, 190–92; flounder, 174; hardyhead, 91; larvivorous, 90–91; mulloway, 126, 174; Pondi, 179, 194; silver bream, 125; tommy ruff, 126; yellow-eye mullet, 126, 174, 191
fisheries, 127–28, 134, 137, 162; and seals, 169–70, 175, 177, 181, 186, 192–93
Fisheries Act 1904, 127–28, 133
Fisheries Act Amendment 1909, 127–28

254 | INDEX

fishing, 4, 10, 13, 21, 101; fishing lobby, 136; and migratory birds, 155, 162; and pelicans, 120–23, 126–31, 133–34, 136–37; and seals, 169–70, 172–78, 181, 184–87, 189–90, 192–93
fishing nets, 4; weaving, 35–36; pelicans attracted to, 128, 130; seals attracted to, 170, 172–74, *174*, 176–77, 185–86, 192
fish kills, 194
Fivebough and Tuckerbil Swamps, 7–8, *8*, 21; and ducks, 97–104, 107, 117–19; and migratory birds, 141, 165–66
Flakelar, Danielle Carney, 5–6, 24–26, 31–35, 38–39, 40–41, 43–44
fleas, 65
flooding, 6, 11–12, 23, 26, 190, 195; and containment of swamps, 46–47, 52, 56–58, 62–64, 71; and disease, 80–81; and ducks, 97, 101–2, 104–5, 115; and pelicans, 125, 139. *See also* rice farming
floodplains, 12, 49, 104, 144, 163, 166; and weaving, 23–25, 39–40, 42
floodways, 100
flounder, 174
flying foxes (fruit bats), 116
Foreign Affairs Department, 159
foxes, 42, 136
Franklin River, 150
Fraser, Malcolm, 162
freshwater, 3–4, 6, 38, 140; and seals, 168–69, 175–76, 186, 190–92, 195; species, 11, 49, 172
Frith, Harry, 102, 114–17, 119, 145, 150–51, 157, 159–61
Frog's Hollow, 56

galahs, 107
game laws, 109–10, 132, 135
Game Protection Act 1866, 110
game reserves, 42, 110, 123, 155
gannets, 127
Gardner, Alex, 164
Garget, John, 59
gender, 9, 182
generations: weaving as, 29–38
Georgia, 108
germ theory, 48, 59, 62–63
Giabal people, 49–50, 52
Gleeson, M. J., 87
Goldsworthy, Simon, 178, 188
Goolwa, 28, 126; Goolwa Barrage, *171*
government conservation, 21, 143–46, 149
Gowrie Creek, 50, 66
grasshoppers, 112

grazing, 13, 84, 101–3, 145; and containment of swamps, 47, 53, 55–57, 66, 69; and weaving, 23–24, *24*, 30, 40–42
great auks, 130
Great Australian Bite, 6
Great Dividing Range, 46, 49
Gregory, Augustus Charles, 56–57
grey teal (*Anas gracilis*), 104–7
Griffith, 113–14
Griffiths, Tom, 54
Groom, William Henry, 53–56, 59
Grosz, Elizabeth, 48
Gunbower Forest, 163

Halliday, Stephen, 53
Hamby, Louise, 36
Hamlyn-Harris, Ronald, 69, 87–91, 94
Hancock, W. K., 54
hardyhead (*Craterocephalus fluviatilis*), 91
Hattah-Kulkune Lakes, 163
haul-out sites, 170, 178, 185
health, human, 21, 52, 54–59, 61–65. *See also* public health
Hera-Singh, Garry, 175–77, 193
herons, 107, 127
Herring, Matthew, 118
Hill, Glen, 173–74, 176–77, 190–92
Hill, Tracy, 173–77, 189–92
Hindmarsh Island, 28–29
histories. *See* environmental history; more-than-human histories
home, the, 21, 91–96
hookworm, 69
horses, 52, 55, 66, 86
horticulture, 83
Hughes, William, 80
Hume Dam, 14
hunting: and ducks, 16, 100, 105–10, 112, 115, 117; and migratory birds, 145–46, 148, 155, 158–59; and pelicans, 36–37, 123, 128, 132–35, 138; by seals, 177. *See also* sealing
Huxley, Sir Julian, 115
hydraulic engineering, 55–56
hydrology, 12, 69
hygiene and cleanliness, 48, 52, 60, 62–63, 87–88

ibis: glossy ibis, 107; white ibis, 71–72, 97, 107
icon sites, 29
imagined ecologies, 119; and disease, 74–75, 77–78, 86, 92, 95–96; and seals, 171; and the wetlands category, 122, 142
immigration, 48, 83, 156, 164

improvement, 20, 48, 53–58
India, 78, 87, 131
Indonesia, 180
industrialization, 144–45
infecting, 20, 73–78; and agriculture, 83–91; and the home, 91–96; and the troubling of boundaries, 80–83; and war, 78–80
Ingold, Tim, 27–28
Inland Rivers Network, 103
insecticide, 94–95. *See also* DDT
insects, 70, 107. *See also* mosquitoes
interdisciplinary approaches, 7, 16–17, 198
International Council for Bird Protection, 146–47
International Union for Conservation of Nature, 149, 178
International Wildfowl Research Bureau, 146–47
interspecies relationships, 19, 74–76, 95
invasive species, 11, 172, 189–91, 197. *See also* European carp
Ipswich, 52, 63
irrigation, 3, 5–7, 13–14, 62, 196; and canalization, 139; Coleambally Irrigation Area, 5; and disease, 73–75, 77, 79–81, 83–84, 86–87, 89, 91–93, 95; and migratory birds, 97, 99–101, 103–4, 111–16; Murray valley irrigation districts, 5, 91, 114; and seals, 169, 172, 180–81, 193; and weaving, 24–25, 35, 38, 40. *See also* Murrumbidgee Irrigation Area
Irrigation Research Extension Committee, 113–14
islands, 169–70, 181–85; in the Bass Strait, 159; in the Coorong lagoon, 120–30, 132–33, 135–40, 123–27. *See also* Hindmarsh Island; Kangaroo Island; Macquarie Island; Pacific Islands; Pelican Islands

Japan, 87, 91, 124, 145
Japan-Australia Migratory Bird Agreement, 15, 21, 141, 143–44, 149–60, 163, 165
Jarowair, 64
Java, 87
Jerome, Paddy (Elder), 64

Kakadu National Park, 151, 163, 166
Kamilaroi, 41
Kangaroo Island, 170, 178, 181–85, *182*
kangaroos, 140
Katinyeri, Aunty Dorrie (Elder), 37
kayi. See spiny sedge
Kearney's Spring Historical Park, 46

kengk muldi. See bush coastal Olearia
Kerang Wetlands, 163
Kerr, Sir John, 162
kidnapping, 180–81
Kinghorn, James, 100, 106–9, 111–14, 117, 119
Kingsford, Richard, 105, 117–19
knowledge. *See* cultural knowledge; women's knowledge
Koltoli. See Latham's snipe
Korea. *See* Republic of Korea
Kukabrank Country, 30
kundui. See sweet appleberry

Lachlan River, 33
lagoons. *See* Coorong lagoon; Pelican Lagoon
lakan. See weaving
Lake Albert, 7, 154; and pelicans, 123–24, 127, 137; and seals, 168–69, *171*, 190
Lake Alexandrina, 3, 7, 14, 36; and pelicans, 123–24, 127, 137; and seals, 168, *171*, 179, 186
Lake Eyre, 125
Lake Mulwala, 14
lakes. *See* Hattah-Kulkune Lakes; Lake Albert; Lake Alexandrina; Lake Eyre; Lake Mulwala; Lower Lakes
Lake Victoria Reservoir, 14
landscape ecology, 118
Land Use Division, 162–63
Langford-Smith, Trevor, 14
larvivorous fish, 90–91
Latham's snipe (*Gallinago hardwickii*), 141, 143, 145, 151–55, 157, 165
law, 54, 82; and ducks, 109–10, 117; and migratory birds, 145, 156, 158; and pelicans, 122–23, 126–27, 131–33, 135, 137; and seals, 178–79, 183; and weaving, 36, 39–40, 43–44
leaking, 19–20, 46–48, 57–58, 60, 71–72
Leeton, 73, 82–83, 86, 88, 92, 102
Limpindjeri, 154
Ling, John, 183–84
lippia, 42
liver fluke, 112–13
Living Murray initiative, 29
lobbying, 39, 41, 53, 55, 103, 129, 136, 155
Local Government Act 1906, 82
long-nosed fur seal (*Arctocephalus forsteri*), 3–4, 21, 170–73, *171*, 177–80, 183–89
Lower Lakes, 28–30, 126, 168–70, 172–73, 175–79, 184–87, 190, 192. *See also* Lake Albert; Lake Alexandrina

Mackay, Alice, 178
Macquarie Island, 183

Macquarie Marshes, 5–7, *8*; and weaving, 23–25, *24*, 27, 31–35, 38, 41–42
Macquarie River, 5, 25, 38, 44
magpie, 125
magpie geese, 151–52
Maitland, 70
malaria, 20, 54, 64, 68–69, 74–78; and agriculture, 83–91; and the home, 91–96; and the troubling of boundaries, 80–83; and war, 78–80
mangrove, 162, *161*
Mansfield-Cameron, Meryl, 29–31, *31*, 35
Mansfield-Cameron, Peter (Elder), 30, 179
manure, 62
marshes. *See* Macquarie Marshes
McClure, Elliott, 157
McKay, Clare, 164
McKeown, Keith C., 85–88, 101–2
measles, 63
Mellor, Doreen, 36
Mellor, John, 129, 135, 137–38
Meningie, 127, 130
methane, 11
MIA. *See* Murrumbidgee Irrigation Area
miasma, 47–48, 53–59, 61, 63, 74, 76
migrating, 20; fish, 170; migratory species, 176, 196; nomadic vs. migratory, 125. *See also* migratory birds
Migratory Animal Pathological Survey, 157
migratory birds, 10, 15–16, 21, 103, 105, 141–44, 196; and the Australian environmental movement, 144–46; and the global environmental crisis, 147; and government conservation, 144; and the Japan-Australia Migratory Bird Agreement, 149–60; and the Ramsar Convention, 146–52; and wetlands survey, 160–67
Migratory Bird Treaty of 1916, 135–36. *See also* Japan-Australia Migratory Bird Agreement
Millennium Drought, 5, 118, 175, 190–92, 196
missions and missionaries, 36, 42, 64, 101
Moggridge, Bradley, 41
more-than-human histories, 7–9, 11, 15–19, 27–28
more-than-human relationships, 9, 15, 17–20, 28, 173, 180, 198
Morgan, Frederick, 83–84
mosquitoes, 9–10, 20–21, 73–78, *79*, 93; and agriculture, 83–91; and containment of swamps, 47–48, 64–72; and the home, 91–96; and migratory birds, 143, 156; and the troubling of boundaries, 80–83; and war and malaria, 78–80, *79*, 93
mountains. *See* Bunya Mountains
mudflats, 98, 102–3
Mudgee, *8*, 27–28, 33
mulloway, 126, 174
multispecies perspective, 7, 10, 17–19, 27–28, 196, 198. *See also* more-than-human histories
Murray Cod, 179, 194
Murray-Darling Basin Wetlands Working Group, 39–40
Murray Lower Darling Rivers Indigenous Nations, 39
Murray River, 3–4, 14, 195; and disease, 73–75, 79–80, 82, 95–96; and migratory birds, 154, 163; and pelicans, 123, 139; and seals, 168–69, 179, 185–86, 190–91; and weaving, 35, 37–38
Murray River Commission, 3
Murray Valley encephalitis, 77, 156
Murray valley irrigation districts, 5, 91, 114
Murrumbidgee Irrigation Area (MIA), 5, *8*, 16, 20–21, 100–2; and disease, 73–78, *75*, 82–84, 86–92, 95–96; and ducks, 97–100, 104, 106–7, 111–15, 118–19; and migratory birds, 151, 165
Murrumbidgee River, 14, 73, 79–80, 82, 97, 101, 104–5
Muruwari, 196
museums, 37–38, 157; Australian Museum, 106–7
myxomatosis, 95

Narranderra, 87
Nash, Linda, 76
National Parks and Wildlife Act 1967, 117
National Parks and Wildlife Act 1972, 138, 178, 184
National Parks and Wildlife Act 1974, 117
National Parks and Wildlife Commission and Service, 150
National Parks and Wildlife Service (NPWS), 41, 43–44, 117, 123
native species, 11, 100, 112, 130–35, 172, 189; Overabundant Native Species Management Plan, 178; pests, 109–11
native title, 29, 39–41
nature, 9–10
Neilley, J. H., 83
neoliberalization, 121
New Guinea, 79, 96, 145, 157–58
New Health Society, 87

New South Wales (NSW), 3, 7, 14–16, 50, 70, 176; and disease, 73–75, 80, 82–83, 92–96; and ducks, 98–101, 107, 109–11, 113, 117, 119; and migratory birds, 141, 145, 151, 157, 160–61, 167; and pelicans, 132, 134; and weaving, 23, 25–28, 33, 35, 41–42

New South Wales Water Conservation and Irrigation Commission (WCIC), 74, 77, 82–85, 87, 90–92, 95

New Zealand, 118, 134, 145–46, 183, 187, 189

New Zealand fur seal (*Arctocephalus forsteri*), 187–89. *See also* long-nosed fur seal

Ngaitji (creation ancestors), 132

Ngarrindjeri, 4–5, 21, 167; and pelicans, 120, 123–26, 132–33, 137–38; and seals, 169–70, 178–79, 181, 185, 187; and weaving, 28–30, 36–37

Ngarrindjeri Regional Authority, 167

Ngurunderi, 179

Nichols, Lettie, 30, 37

nomadism, 155–56

nori. *See* pelicans

North America, 99, 157

Northern Territory, 32, 35–36, 149, 151, 154, 191

NPWS. *See* National Parks and Wildlife Service

NSW. *See* New South Wales

NSW Fauna Protection Act 1948, 117

Olsen, Penny, 133

opportunism, 155–56, 177–78, 185

ornithology, 101, 112; and migratory birds, 145–46, 153–57; and pelicans, 120–23, 126, 129–31, 134–37, 139

Overabundant Native Species Management Plan, 178

Oxley Creek, 55

Pacific black duck (*Anas superciliosa*), 104–7

Pacific Islands, 156, 158–59

parasites, 64, 69, 74–76, 78, 80–81, 95

pastoralists, 25, 42, 44, 49–50, 110, 112–13, 115

Pastures and Stock Protection Act, 109

Pelican Islands, 125–30, 135

Pelican Lagoon, 125–26

pelicans, 120–23, 138–40, *124*; and carp, 192; and the Coorong, 123–27, *121*; protection of, 135–38; slaughter of, 127–35; and seals, 177, 179; and weavers, 36–37

pelts. *See* skins and pelts

pest control model, 116–17, 151

pests, 16, 65, 68–70, 132; ducks as, 99–100, 105–9, 112–13, 115–17, 119; insects as, 82, 89, 95, 134, 155; migratory birds as, 151, 155; native, 109–11; pelicans as, 120, 126–29; and protection, 133–35; seals as, 172, 190–91

pheasants, 110, 131

phytoplankton, 169

pigs, 60–62, 83

Pine River, 55

pink-eared duck (*Malacorhynchus membranaceus*), 104–5, 107

pipis, 4, 169, 193

Pittock, Jamie, 103, 164

play behavior, 177

Point McLeay Mission, 36

policies, 20, 33, 39–40, 42–43, 103, 158, 162

politics: and disease, 74, 78–80, 83, 96; and ducks, 99, 105, 118; of knowledge, 77; and migratory birds, 144–45, 149–52, 154, 158, 163–64; multiscalar, 142–43, 167; multispecies, 27; and pelicans, 132, 136–37; postcolonial, 27; and weaving, 28–29, 36. *See also* water politics

Pondi (Murray Cod), 179, 194

Port Essington, 78

postcolonial, 13, 27

Price, Dr. Thomas, 83, 85–87

private property, 21, 121, 123, 131–37

privatization, 133; of protection, 135–38. *See also* private property

protection and protected areas, 5, 21, 43; and ducks, 116–17; and migratory birds, 149–50, 158, 166–67; and pelicans, 120–24, 126–38, 140; privatization of, 135–38

public health, 59, 63–65, 67–70, 76–78, 82–86, 91–92, 95–96

quail, 108

Queensland, 7, 20, 32, 36, 116, 145; and containment of swamps, 46, 50–57, 62–65, 67–70

Queensland Government, 53, 56–57, 64, 68, 70

Queensland Rat and Mosquito Prevention and Destruction Act of 1916, 67

Queens Park, 55, 60

Quirindi, 70

rabbits, 30, 95, 131, 136

race and racial ideologies, 9, 18–19, 43, 197; and containment of swamps, 47–48, 52, 59, 62, 71; and disease, 78, 87–89; and

migratory birds, 153, 156; and pelicans, 137–38, 140
racism. *See* race and racial ideologies
rainfall, 35, 49, 68, 85, 154–55, 168, 196
Ramsar Convention on Wetlands of International Importance (1971), 7, 12, 15, 21, 29, 42; and ducks, 100, 102–3, 117–18; and migratory birds, 124, 141, 143–44, 146–52, 160, 163–67
Ramsar Wetland of International Importance, 3, 5, 103
Ratcliffe, Francis, 114–16
rats, 65, 76
Raukkan, 36–37
recession, 162
reed beds, 5, 23, 25, 38, 195
reefs, 148, 164
Reid, Julian, 124–25, 139
relationality, 18–19, 75–76, 154
Republic of Korea, 163
rice farming, 5, 14, 16, 21, 101–2; and disease, 86–89, 91; and ducks, 97–100, 98, 104–15; and migratory birds, 151, 155, 165; and postwar ecology and conservation science, 115–19
Ricegrowers' Association of Australia, 113–14, 118
Rice Marketing Board, 87
rights, 27, 39–40; hunting rights, 159; land rights, 13, 42, 138, 148–49, 158–59; native title rights, 29; rights to Country, 39, 43; water rights, 13, 196
rippling: capitalism, 21, 169, 172–73, 180–81; of sealing, 172, 187, 191–93
river red gum, 23, 25, 49, 166, 195
rivers. *See* Alligator River; Franklin River; Lachlan River; Macquarie River; Murray River; Murrumbidgee River; Pine River; Swan River
Roberts, Frederick Hugh Sherston, 95
Robin, Libby, 116, 153, 155–56
Rogers, Tracey, 177
rookeries, 21, 123–27, 129–30, 132, 136, 139. *See also* pelicans
Ross, Ronald, 78, 94
Royal Australian Ornithological Union, 101, 136
Royal Zoological Society, 111–12
Russell, Lynnette, 182
Rutherford, John, 14
Ruwe/Ruwar (lands, waters and all living things), 169
Ryan, Lyndall, 182

salinity, 3–4, 29, 73, 83, 139, 169, 175; salinization, 38, 87
salt water, 168–69, 175, 186
sanctuaries, 102, 123, 130, 135, 151, 158
Sandlos, John, 186–87
scarlet fever, 53
sciences, 9, 59–60, 119, 143, 153, 198. *See also specific sciences*
sealing, 21, 159, 171–73, 178–87, 191–93
sea lions, 170, 181, 184, 188. *See also* Australian sea lion
seals, 10; Australian fur seal (*Arctocephalus pusillus*), 170, 188–89; and belonging, 187–93; and capitalism, 173–81; long-nosed fur seal (*Arctocephalus forsteri*), 3–4, 21, 170–73, *171*, 177–80, 183–89; New Zealand fur seal (*Arctocephalus forsteri*), 187–89; numbers affected by violence, 181–87; Tasmanian fur seal (*Arctocephalus tasmanicus*), 188
Serventy, D. L., 159
sewage, 58, 60, 65–66, 119
sewage treatment works, 14, 98, 102, 165
Shaughnessy, Peter, 178, 188
sheep, 13, 30, 42, 49, 52, 83, 110, 113
Shillingsworth, Bruce, 196
silver bream, 125
silver gulls, 127
skins and pelts, 132, 134, 181, 183–84, 188
smallpox, 126
smoking bush. *See* bush coastal Olearia
snails, 112, 134
Snowy Mountains Hydro-Electric Scheme, 14, 97, 113
SOA. *See* South Australian Ornithological Association
Soulé, Michael, 147
South Australia, 3, 6–7, 21, 27–28, 30, 80–82; and pelicans, 120, 126–27, 129, 132–33, 136, 139; and seals, 168, 178–79, 181–84, 189. *See also specific locations*
South Australian Animals Protection Act 1912, 135
South Australian Ornithological Association (SOA), 129, 131, 134–37, 139
Southeast Asia Treaty Organization, 157
Southern Ocean, 3, 168, 193
sparrows, 107
spiny sedge (*Cyperus gymnocaulos*), 26, 28, 29
spirituality, 13, 26, 28, 34, 36, 39–40, 179
spoonbills, 107
State Records of South Australia, 122

Stegomyia mosquito (*Stegomyia fasciata*), 65–69. *See also* mosquitoes
Stoddard, Herbert, 108
storm surges, 11–12, 195
Sturt, Charles, 37, 179
subsistence hunting, 159. *See also* hunting
Sumner, Darrell, 179
swamps. *See* Barren Box Swamp; Fivebough and Tuckerbil Swamps; Toowoomba swamps; underground swamps
Swann, Hugh, 55–56
Swan River, 136
sweet appleberry, 29
Sydney Morning Herald, 87, 101, 111–12
Syria, 79

Takasuka, Isaburo, 86
Tanganalun. *See* Tanganekald
Tanganekald, 29–30, 126
Tasmania, 145, 150, 152, 155, 159, 181
Tasmanian fur seal (*Arctocephalus tasmanicus*), 188
Taylor, Frank H., 75, 80, 82–83, 92–95
technologies, 9–11, 62, 93–94, 131
Temperumindjeri, 30
terns, 127
Thomson M. B., John, 59
Thorneloe-Smith, James, 57–59
tommy ruff (Australian herring), 126
Toowoomba, city of, 20, *51*, 46–53; and cesspits, 58–60; and Chinese market gardeners, 60–64; and improvement, 54–58; and miasma, 53–54, 56; and mosquitoes, 64–72; and stinky swamps, 54, 56; and underground swamps, 58–60; and wells, 58–60, 63
Toowoomba Council, 47–48, *47*, 53–60, 62–65, 67–68, 70
Toowoomba swamps, 7, 8, 48, *48*, 50–54, *51*, 56
tortoises, 128
traps, 30, 36
treaties, 124, 135, 149–50, 163–64; Migratory Bird Treaty of 1916, 135; migratory bird treaty with Japan, 141, 150, 152, 158, 160
Tregilges, W., 130–31, 137–38
Trevorrow, Ellen, 37
Trevorrow, Tom, 37
Trunkeena, 30
Tsing, Anna, 75–76, 173, 180, 193
Tunstall, Elizabeth Dori, 31
turtles, 128
typhoid, 53–55, 58–61, 63–64

UK, 148–49. *See also* Britain; England
underground swamps, 58–64
United States, 79, 83, 116, 186, 196; and migratory birds, 147–48, 152, 156–57, 162; and pelicans, 129, 131, 134–35. *See also* California; Georgia
urbanization, 117, 145
US Army, 157
USSR, 148–49

Vaughan, Crawford, 135
Victoria, 3, 15, 80; and carp, 190; and ducks, 110–11; and migratory birds, 145, 155; and weaving, 35–36
violence, 42, 49–50, 171, 182

Wailwan, 5–6, 20, 23–27, 38–44
Wakka Wakka, 24
Wakool, 91–92
warming, 195
wastelands, 10, 13, 48, 54
Water Act 2007, 164
waterbirds, 10, 15, 21, 25, 49–50, 125, 177; Australian Committee on Waterbirds, 160; and boundaries, 100, 102, 107, 110, 117; and migration, 143, 146–48, 152, 161–62; A Survey of Wetland Habitats of Australian Waterbirds, 160; Toowoomba Waterbird Habitat, 71–72. *See also specific species*
water couch, 102
water diversion, 5–6, 11–16, 25, 35, 40, 87, 99–100, 165. *See also* drainage; irrigation
water hens, 5, 97
water management, 25–26, 35, 38, 41, 73, 143, 164, 195–96
water politics, 16, 20, 24–27; weaving as, 38–41
water shortages, 119, 196
water siphoning, 24, 40
water storage, 15, 65, 101, 146
water tables, 59, 87
Watson, Irene, 133
Wayilwan. *See* Wailwan
WCIC. *See* New South Wales Water Conservation and Irrigation Commission
weaver birds, 27–28
weaving, 6, 10, 20, 23–29; as access, 41–45; as generations and Country, 29–38; as water politics, 38–41
Webb, Cath, 103
weeds, 42, 66, 86, 104, 106–7, 109, 134

Weilwan. *See* Wailwan
Weir, Jessica, 167
wells, 14, 46, 52, 58–64, 67, 72
well water, 52–53, 58, 60, 63, 66
West Creek, 46, 71
Western Aboriginal Women's Network, 34
wetlands, 12–14; as a category, 6–7, 12, 15, 141–43, 146–48, 160, 164–65, 197
wetlands ecology, 147
wetlands surveys, 21, 143, 160–67
wet meadows, 161
whistling duck (plumed-tree duck, *Dendrocygna arctuata*), 104–5
White, Samuel, 101, 129–30
White, Steven, 134–37
Whitlam, Gough/Whitlam government, 144, 149–50, 152, 162
Wiersma, Yolanda, 186
wilderness, 148–50, 159, 167
Wildlife Conservation report, 144, 146

Wildlife Survey Section, 114, 116
Wiradjuri, 28, 32–33, 42, 101, 105, 113
women's knowledge, 26–29, 32–34, 37
World Heritage Convention, 150
world-making, 27–28, 74–75, 96
world wars, 20, 77, 96, 156; World War I (Great War), 68, 74, 79, 81, 92, 111; World War II, 70, 74, 76, 79, 91, 93, 105, 113, 116, 142
World Wildlife Fund, 103

Yamashina, Yoshimaro, 145
Yanyuwa, 154
Yaraldi, 154
Yarkuwa Indigenous Knowledge Centre Aboriginal Corporation, 167
yellow-eye mullet, 126, 174, 191
yellow fever, 65, 69

zoology, 106–7, 111–12, 115–16, 119, 188

WEYERHAEUSER ENVIRONMENTAL BOOKS

Wetlands in a Dry Land: More-Than-Human Histories of Australia's Murray-Darling Basin, by Emily O'Gorman

Seeds of Control: Japan's Empire of Forestry in Colonial Korea, by David Fedman

Fir and Empire: The Transformation of Forests in Early Modern China, by Ian M. Miller

Communist Pigs: An Animal History of East Germany's Rise and Fall, by Thomas Fleischman

Footprints of War: Militarized Landscapes in Vietnam, by David Biggs

Cultivating Nature: The Conservation of a Valencian Working Landscape, by Sarah R. Hamilton

Bringing Whales Ashore: Oceans and the Environment of Early Modern Japan, by Jakobina K. Arch

The Organic Profit: Rodale and the Making of Marketplace Environmentalism, by Andrew N. Case

Seismic City: An Environmental History of San Francisco's 1906 Earthquake, by Joanna L. Dyl

Smell Detectives: An Olfactory History of Nineteenth-Century Urban America, by Melanie A. Kiechle

Defending Giants: The Redwood Wars and the Transformation of American Environmental Politics, by Darren Frederick Speece

The City Is More Than Human: An Animal History of Seattle, by Frederick L. Brown

Wilderburbs: Communities on Nature's Edge, by Lincoln Bramwell

How to Read the American West: A Field Guide, by William Wyckoff

Behind the Curve: Science and the Politics of Global Warming, by Joshua P. Howe

Whales and Nations: Environmental Diplomacy on the High Seas, by Kurkpatrick Dorsey

Loving Nature, Fearing the State: Environmentalism and Antigovernment Politics before Reagan, by Brian Allen Drake

Pests in the City: Flies, Bedbugs, Cockroaches, and Rats, by Dawn Day Biehler

Tangled Roots: The Appalachian Trail and American Environmental Politics, by Sarah Mittlefehldt

Vacationland: Tourism and Environment in the Colorado High Country, by William Philpott

Car Country: An Environmental History, by Christopher W. Wells

Nature Next Door: Cities and Trees in the American Northeast, by Ellen Stroud

Pumpkin: The Curious History of an American Icon, by Cindy Ott

The Promise of Wilderness: American Environmental Politics since 1964, by James Morton Turner

The Republic of Nature: An Environmental History of the United States, by Mark Fiege

A Storied Wilderness: Rewilding the Apostle Islands, by James W. Feldman

Iceland Imagined: Nature, Culture, and Storytelling in the North Atlantic, by Karen Oslund

Quagmire: Nation-Building and Nature in the Mekong Delta, by David Biggs

Seeking Refuge: Birds and Landscapes of the Pacific Flyway, by Robert M. Wilson

Toxic Archipelago: A History of Industrial Disease in Japan, by Brett L. Walker

Dreaming of Sheep in Navajo Country, by Marsha L. Weisiger

Shaping the Shoreline: Fisheries and Tourism on the Monterey Coast, by Connie Y. Chiang

The Fishermen's Frontier: People and Salmon in Southeast Alaska, by David F. Arnold

Making Mountains: New York City and the Catskills, by David Stradling

Plowed Under: Agriculture and Environment in the Palouse, by Andrew P. Duffin

The Country in the City: The Greening of the San Francisco Bay Area, by Richard A. Walker

Native Seattle: Histories from the Crossing-Over Place, by Coll Thrush

Drawing Lines in the Forest: Creating Wilderness Areas in the Pacific Northwest, by Kevin R. Marsh

Public Power, Private Dams: The Hells Canyon High Dam Controversy, by Karl Boyd Brooks

Windshield Wilderness: Cars, Roads, and Nature in Washington's National Parks, by David Louter

On the Road Again: Montana's Changing Landscape, by William Wyckoff

Wilderness Forever: Howard Zahniser and the Path to the Wilderness Act, by Mark Harvey

The Lost Wolves of Japan, by Brett L. Walker

Landscapes of Conflict: The Oregon Story, 1940–2000, by William G. Robbins

Faith in Nature: Environmentalism as Religious Quest, by Thomas R. Dunlap

The Nature of Gold: An Environmental History of the Klondike Gold Rush, by Kathryn Morse

Where Land and Water Meet: A Western Landscape Transformed, by Nancy Langston

The Rhine: An Eco-Biography, 1815–2000, by Mark Cioc

Driven Wild: How the Fight against Automobiles Launched the Modern Wilderness Movement, by Paul S. Sutter

George Perkins Marsh: Prophet of Conservation, by David Lowenthal

Making Salmon: An Environmental History of the Northwest Fisheries Crisis, by Joseph E. Taylor III

Irrigated Eden: The Making of an Agricultural Landscape in the American West, by Mark Fiege

The Dawn of Conservation Diplomacy: U.S.-Canadian Wildlife Protection Treaties in the Progressive Era, by Kirkpatrick Dorsey

Landscapes of Promise: The Oregon Story, 1800–1940, by William G. Robbins

Forest Dreams, Forest Nightmares: The Paradox of Old Growth in the Inland West, by Nancy Langston

The Natural History of Puget Sound Country, by Arthur R. Kruckeberg

WEYERHAEUSER ENVIRONMENTAL CLASSICS

Environmental Justice in Postwar America: A Documentary Reader, edited by Christopher W. Wells

Making Climate Change History: Documents from Global Warming's Past, edited by Joshua P. Howe

Nuclear Reactions: Documenting American Encounters with Nuclear Energy, edited by James W. Feldman

The Wilderness Writings of Howard Zahniser, edited by Mark Harvey

The Environmental Moment: 1968–1972, edited by David Stradling

Reel Nature: America's Romance with Wildlife on Film, by Gregg Mitman

DDT, Silent Spring, and the Rise of Environmentalism, edited by Thomas R. Dunlap

Conservation in the Progressive Era: Classic Texts, edited by David Stradling

Man and Nature: Or, Physical Geography as Modified by Human Action, by George Perkins Marsh

A Symbol of Wilderness: Echo Park and the American Conservation Movement, by Mark W. T. Harvey

Tutira: The Story of a New Zealand Sheep Station, by Herbert Guthrie-Smith

Mountain Gloom and Mountain Glory: The Development of the Aesthetics of the Infinite, by Marjorie Hope Nicolson

The Great Columbia Plain: A Historical Geography, 1805–1910, by Donald W. Meinig

CYCLE OF FIRE

Fire: A Brief History, second edition, by Stephen J. Pyne

The Ice: A Journey to Antarctica, by Stephen J. Pyne

Burning Bush: A Fire History of Australia, by Stephen J. Pyne

Fire in America: A Cultural History of Wildland and Rural Fire, by Stephen J. Pyne

Vestal Fire: An Environmental History, Told through Fire, of Europe and Europe's Encounter with the World, by Stephen J. Pyne

World Fire: The Culture of Fire on Earth, by Stephen J. Pyne

ALSO AVAILABLE:

Awful Splendour: A Fire History of Canada, by Stephen J. Pyne